ENVIRONMENT AND ECOLOGY

ENVIRONMENT AND ECOLOGY

(As per WBUT Syllabus)

Anil Kumar De

Formerly Professor and Head
Department of Chemistry
&
Dean, Faculty of Science
Visva Bharati University, Santiniketan
West Bengal

Arnab Kumar De

Reader in Chemistry
Raja Peary Mohan College, Uttarpara
West Bengal

NEW AGE
WBUT

An Imprint of

NEW AGE INTERNATIONAL (P) LIMITED, PUBLISHERS

New Delhi • Bangalore • Chennai • Cochin • Guwahati • Hyderabad
Jalandhar • Kolkata • Lucknow • Mumbai • Ranchi
Visit us at **www.newagepublishers.com**

Published by New Age International (P) Ltd., Publishers
First Edition: 2010

Branches:

- 36, Malikarjuna Temple Street, Opp. ICWA, Basavanagudi, **Bangalore**. ✆ (080) 26677815
- 26, Damodaran Street, T. Nagar, **Chennai**. ✆ (044) 24353401
- CE-39/1016, Carrier Station Road, Near Paulson Park Hotel, Ernakulam South, **Cochin.** ✆ (0484) 4051303
- Hemsen Complex, Mohd. Shah Road, Paltan Bazar, Near Starline Hotel, **Guwahati**. ✆ (0361) 2543669
- No. 105, 1st Floor, Madhiray Kaveri Tower, 3-2-19, Azam Jahi Road, Nimboliadda, **Hyderabad**. ✆ (040) 24652456
- RDB Chambers (Formerly Lotus Cinema) 106A,1st Floor, S.N. Banerjee Road, **Kolkata**. ✆ (033) 22275247
- 16-A, Jopling Road, **Lucknow**. ✆ (0522) 2209578
- 142C, Victor House, Ground Floor, N.M. Joshi Marg, Lower Parel, **Mumbai**. ✆ (022) 24927869
- 22, Golden House, Daryaganj, **New Delhi**. ✆ (011) 23262370, 23262368

ISBN : 978-81-224-2751-6

Rs. 90.00

C-09-06-3701

Printed in India at Ram Printograph, Delhi.
Typeset at In-house.

PUBLISHING FOR ONE WORLD
NEW AGE INTERNATIONAL (P) LIMITED, PUBLISHERS
4835/24, Ansari Road, Daryaganj, New Delhi-110002
Visit us at **www.newagepublishers.com**

Preface

The environment courses for various disciplines (Arts, Science, Commerce, Engineering, etc.) were introduced in Indian Universities, IITS and Technical Universities in 2003 in the context of the landmark judgements of the Hon'ble Supreme Court of India in 1991 and again in 2003. This raised clouds of confusion among the teachers and students of various disciplines due to lack of orientation among the teachers and also lack of suitable textbooks about the subject.

The authors with their long experience in environmental teaching and research have taken up the challenge to fulfil the need of the hour by publishing series of books for this purpose since 2001. Now they have launched this project of Environment & Ecology for B.E. semester-II course of W.B. University of Technology. The book has been designed in terms of the relevant syllabus and presented in logical sequence in seven Chapters—General (Introduction), Population and Resources, Elements of Ecology, Air Pollution and Control, Water Pollution and Control, Land Pollution and Noise Pollution. In order to maintain the logical sequence of the subject matter, overlapping of contents has been avoided and some additional matter included in some chapters (chapters 4 and 5) to give proper perspectives to the subject which will be helpful to the students in their future career. The Feedback Exercises, Glossary and Sample Question Papers will help to reinforce the understanding of the subject by the students. The Author's mission is to motivate the engineering students with environmental culture.

Constructive suggestions are invited from engineering students and teachers for further improvement of the book in its future editions. It is a pleasure to place on record our sincere thanks to Mr. S. Gupta, *Managing Director* and V.R. Damodaran, *Chief Editor*, New Age International (P) Ltd., New Delhi for prompt publication and to Mrs. Kalpana De, Mrs. Chandrima and Charchita Datta for constant encouragement.

P 27/1 CIT Sch. VII M
'Udichi' Kolkata-700054

Anil Kumar De
Arnab Kumar De

Syllabus

ENVIRONMENT & ECOLOGY (W.B.U.T. SYLLABUS)

Code: Ch 101
Contacts: 3L = 3
Credits: 3

General

Basic ideas of environment, basic concepts related to environmental perspective, man, society and environment, their inter relationship. 1L

Mathematics of population growth and associated problems, definition of resource, types of resource, renewable, non-renewable, potentially renewable, effect of excessive use vis-à-vis population growth, definition of pollutant and contaminant. Environmental impact assessment. 2L

Environmental degradation: Acid rain, toxic element, particulates, noise pollution, air pollution and its effect on man. 1L

Overall methods for pollution prevention, environmental problems and sustainable development, components of environment. 1L

Ecology

Elements of Ecology: System, open system, closed system, definition of ecology, species, population, community, definition of ecosystem, biotic and abiotic components.

Ecological balance and consequence of change: Effect of abiotic factor on population, flow chart of different cycles with only elementary reaction [oxygen, nitrogen, phosphate, sulphur], food chain [definition and one example of each food chain] 3L

Air Pollution and Control

Atmospheric Composition: Troposphere, stratosphere, mesosphere, thermosphere, tropopause, stratopause and mesopause. 1L

Energy Balance: Conductive and convective heat transfer, radiation heat transfer, simple global temperature modal [Earth as a black body, earth albedo], problems. 3L

Green-house effects: Definition, impact of greenhouse gases on the global climate and consequently on sea water level, agriculture and marine food. 1L

Climate, weather: Difference between climate and weather

Global warming and its consequence: 2L

Adiabatic lapse rate, atmospheric stability, temperature inversion, radiation inversion atmospheric dispersion: Maximum mixing depth, ventilation coefficient, smokestack plumes and atmospheric lapse rate. 1L

The point-source Gaussian plume model excluded.

Source and effect of pollutants: Toxic chemicals in the environment, toxic chemicals in air, suspended particulate matter, carbon dioxide, sulphur dioxide, nitric oxide, lead, carbon monoxide. 2L

Primary and secondary pollutants: Emission standard, criteria pollutant, oxides of carbon, oxide of nitrogen, oxide of sulphur, particulate, PAN

Depletion Ozone layer: CFC, destruction of ozone lair by CFC, impact of other greenhouse gases, effect of ozone modification. 1L

Standards and control measures: Industrial, commercial and residential air quality standard, Control measure (ESP, Cyclone separator, bag house, catalytic converter, scrubber (ventury). Statement with brief reference) 1L

Water Pollution and Control

Hydrosphere: Hydrological cycle 1L

Natural water

Pollutants: their origin and effects: Oxygen demanding wastes, pathogens, nutrients, salts, thermal application, heavy metals, pesticides, volatile organic compounds 1L

River/lake/ground water pollution

River

DO, 5 day BOD test, BOD reaction rate constants, temperature dependents of BOD, effect of oxygen demanding wastes on river [Deoxygenation, reaeration], COD, Oil, Grease, pH. 2L

Lake: Eutrophication [Definition, source and effect] 1L

Ground Water: Aquifers, hydraulic gradient, ground water flow. (Definition only) 1L

Standard and control: Waste water standard [BOD, COD, Oil, Grease], Water treatment system [coagulation and flocculation, sedimentation and filtration, disinfection, hardness and alkalinity, softening], waste water treatment, primary treatment, secondary treatments [Trickling filters, rotating biological contractor, activated sludge, sludge treatment, oxidation ponds], tertiary treatment definition. 3L

Arsenic pollution: Biochemical effect, contamination, speciation 2L

Land Pollution

Lithosphere

Composition

Pollutants

Municipal, industrial, commercial, agricultural, hazardous solid wastes 1L

Recovery and conversion method 2L

Waste and waste management

Land filling, incineration, composting

Noise Pollution 2L

Sources, effects

Definition of noise, effect of noise pollution, noise classification, transport noise, occupational noise, neighbourhood noise, definition of noise intensity, noise threshold limit value.

Contents

1 General

1.1 DEFINITION

What is Environment? The word '**Environment**' originates from 'Environ' which means *"things that surround"*. So Environment implies *all that environs (surrounds) us*, i.e., all the physical and biological surroundings–earth, air, water, men, plants, trees, animals, micro-organisms, buildings, road, etc. The living organisms vary from the lowest micro-organisms such as bacteria, virus, fungus, etc. to the highest including man. Each organism has its own environment (physical and biological). *In other words, Environment is the sum total of all conditions and influences that affect the development and life of all organisms on earth.* Environment has multi-dimensional aspects so that its perception varies from man to man. To some, it is a scenic landscape; to others, it is natural resources (air, water, soil, forests) or industrial pollution, etc. Environment offers different services or benefits to man: water for drinking, air for breathing, agricultural crops for food, medicines from plants, minerals for industries, lakes for recreation and aesthetics, etc. However, environment loses its ability to discharge these functions properly due to stress from man-made activities.

1.2 COMPONENTS OF ENVIRONMENT–INTERRELATIONSHIPS

The environment consists broadly of two components—Non-living or Physical and Living or Biological.

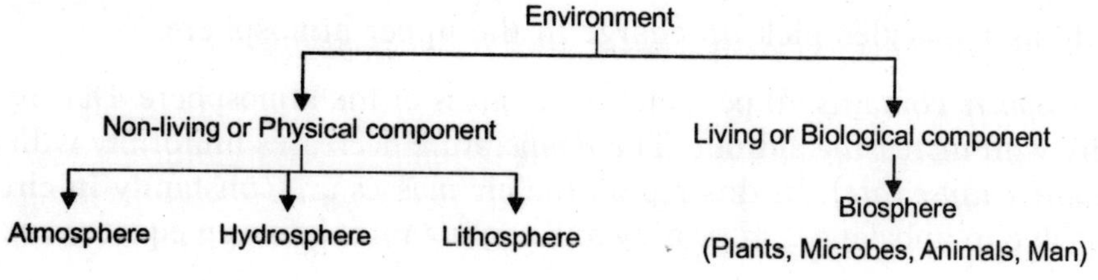

Within an ecosystem (unit of Biosphere which coexists with physical environment) there are dynamic interrelationships between the living forms and their physical Environment. These are manifested by natural cycles which provide a continuous circulation of essential nutrients (N, P, S, C, O) necessary for life in the biosphere. The natural cycles and ecosystems operate in a balanced manner which stabilises the entire biosphere and sustain life processes on Earth.

(1) Atmosphere: The atmosphere has three major constituents—major, minor and trace. Pure (i.e., pollution-free) dry air at ground level has the following components, expressed in *percentage by volume* (within brackets):

Major components:	Nitrogen (78.09)	Oxygen (20.94)	Water vapour (0.1)
Minor components:	Argon (0.9)	Carbon dioxide (0.032)	
Trace components:	Neon (0.0018)	Helium (0.0005)	Methane (0.0002) etc.

The properties of the atmosphere vary much with altitude. The density shows sharp decrease with increasing altitude. Pressure drops from 1 atmosphere at sea level to 3×10^{-7} atmosphere at 100 km above sea level while temperature varies from –92° to 1200° C. The total mass of the atmosphere is about 5×10^{7} tons which is roughly one millionth of the earth's total mass (5×10^{24} kg).

Structure: The atmosphere may be broadly divided into four regions as shown in Table 1.1 below. It extends up to 500 km, with temperature varying from a minimum of –92°C to a maximum of 1200°C.

Table 1.1: Major Regions of the Atmosphere

Region	***Altitude range km***	***Temperature range, °C***	***Important chemical species***
Troposphere	0–11	15 to –56	Nitrogen, Oxygen, Water vapour, Carbon dioxide
Stratosphere	11–50	–56 to –2	Ozone
Mesosphere	50–85	–2 to –92	Oxygen^{+}, Nitric oxide
Thermosphere	85–500	–92 to 1200	Oxygen^{+}, Nitric oxide^{+}

Oxygen^{+} = Oxygen atom with positive charge.

Nitrogen oxide^{+} = Nitric oxide molecule with positive charge.

These atoms/molecules pick up charge in the upper atmosphere.

The *Troposphere* contains 70 per cent of the mass of the atmosphere. Density decreases exponentially with increasing altitude. The temperature decreases uniformly with increasing altitude (*negative lapse rate*). In this region the air masses are constantly in circulation as energy flows due to imbalances in heating and cooling rates between equator and the poles.

In contrast to the Troposphere, the *Stratosphere* is the quiet layer having a *positive lapse rate*, i.e., increase in temperature with increase in altitude. Here ozone is the important species which absorbs ultra violet radiation and raises the temperature. The *ozone layer serves as a protective shield for life forms on earth* from the harmful effects of the sun's ultra violet radiation. Because of slow mixing in the stratosphere, the molecules or particles in the region stay for a long time. If some pollutants enter this layer, they will stay for a long period and slowly come down to the Troposphere, causing long-term global hazards. Introduction of nitrogen oxides by jet planes and refrigerant gases (chlorofluorocarbons) leads to thinning of ozone layer and generates ozone hole.

In the *Mesosphere* temperature decreases with increase in altitude. This is due to low levels of ultra violet species, i.e., ozone.

In the *Thermosphere* temperature rises once again giving a *positive lapse rate*. Here oxygen and nitric oxide ionise after absorption of solar radiation in the far ultraviolet region.

(2) Hydrosphere: The history of human civilisation shows that water supply and civilisation are intimately linked with each other. Several cities and civilisations have vanished due to shortage of water supply.

The world's total quantum of water is 1.4 billion km^3. If all the seabeds could be filled up and brought at the level of the earth's surface, then the entire water in the seas would cover the earth's surface and make it 2.5 km deep water mass. About 97 per cent of the earth's water supply is in the ocean which is unfit for human consumption and other uses due to high salt content. Of the remaining 3 per cent, 2.3 per cent is locked in the polar ice caps and hence out of bounds. The balance 0.7 per cent is available as fresh water but the bulk of it is ground water (0.66 per cent) and the rest 0.03 per cent is fresh water in rivers, lakes and streams. The break-up of this 0.03 per cent is: lakes and ponds, 0.01 per cent, water vapour 0.001 per cent, rivers 0.003 per cent and water confined in plants, animals and chemicals 0.0187 per cent (United Nations Water Conference Report, Argentina, 1977).

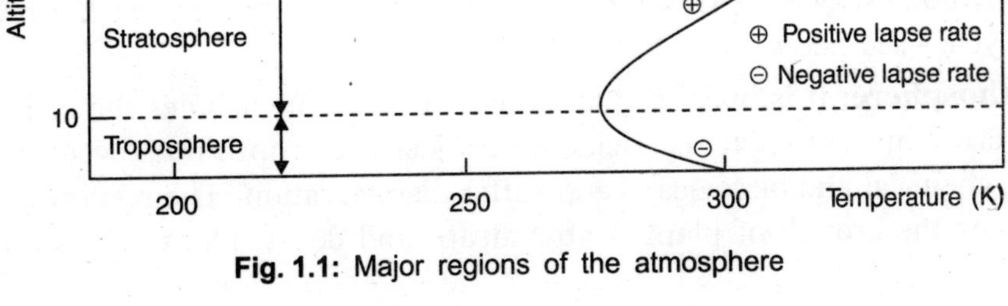

Fig. 1.1: Major regions of the atmosphere

Thus we see that we have a very limited stock of usable water, 0.03 per cent surface water (rivers, streams and ponds)and 0.66 per cent ground water. The quantity of water vapour arising from evaporation of sea water and river water returns by the same volume to the earth's surface by rainfall and back to the water sources. This natural hydrological cycle is more or less balanced in terms of charge (cloud formation) and discharge (rainfall) and is the source of fresh water supply. But we are drawing large quantities of ground water for agriculture and industries while the waste water from these is much polluted and on mixing with rivers is polluting the rivers also.

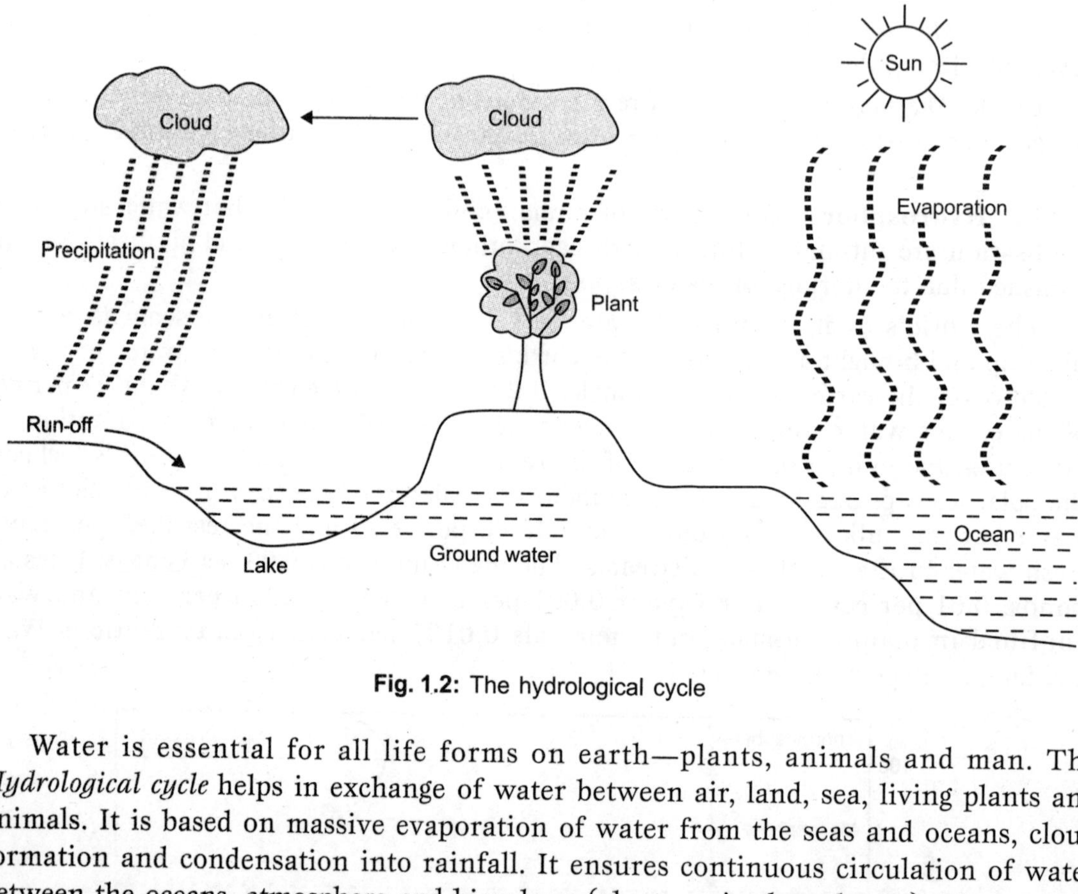

Fig. 1.2: The hydrological cycle

Water is essential for all life forms on earth—plants, animals and man. The *Hydrological cycle* helps in exchange of water between air, land, sea, living plants and animals. It is based on massive evaporation of water from the seas and oceans, cloud formation and condensation into rainfall. It ensures continuous circulation of water between the oceans, atmosphere and biosphere (plants, animals and man). Thus we get our supply and reserves of fresh water. Figure 1.2 illustrates the hydrological cycle (Natural Cycle—see Sec. 2.10).

(3) Lithosphere: It is made of the mantle of rocks. It includes the soil which covers the rock's crust in many places. Rocks are subjected to continuous weathering forces—rain, wind, chemical and biological—and suffer disintegration. The resulting primitive soil is suitable for the growth of plants—after death and decay, plant debris returns to soil.

The mineral component of soil comes from the parent rocks by weathering processes while the organic component is due to plant biomass as well as populations of bacteria, fungi and insects (earthworms). A typical soil, suitable for agriculture, contains about 5 per cent organic matter and 95 per cent inorganic matter. Soil plays an important role as it produces food for man and animals. Good soil and good agriculture are valuable assets for a nation.

In general, soil has a loose structure consisting of solid mineral, organic matter and air space. It shows broadly three zones as its depth increases. The top layer, up to several inches thick, is known as top soil which is an index of soil quality. This is the layer of maximum biological productivity and it contains bulk of organic matter. Hence, it is very important for vegetation cover and agricultural crops. Reckless deforestation causes loss of top soil which means loss of agricultural production. The underlying layer is the subsoil which receives organic matter, salts and clay particles leached from the top soil. The third layer (zone) consists of weathered parent rocks from which soil was formed. Plants draw water and nutrients from soil—they transport water into the plant body (roots and leaves) and discharge excess water into the atmosphere through leaves through the process of transpiration.

The various land forms of the lithosphere are (i) mountain, (ii) plateau and (iii) plain. Mountains provide natural frontier, shelter for tribals, important flora and fauna. They contain forests which are important natural resources. Plateaus are rich in forest and mineral wealth while the plains account for maximum world population because of the convenience for cultivation, communication, transport and industrial growth.

(4) Biosphere: Broadly speaking, the biosphere consists of the earth's crust, hydrosphere, atmosphere and various living species (micro-organisms to man) which exist in the zone 600 metres above earth's surface and 10,000 metres below sea level. Both biosphere and environment have close interactions with each other. Thus oxygen and carbon dioxide level of atmosphere depend on the plant world. Green plants are responsible for accumulation of oxygen in the atmosphere through photosynthesis and decay. In primitive stage the atmosphere was devoid of oxygen and there was no life form on earth. In general, the biosphere is closely related to energy flows in the environment and water chemistry.

1.3 MAN-ENVIRONMENT—INTERACTION

Man is at the centre of the biosphere in dynamic equilibrium with other segments of the environment—air, water and land. His daily life is heavily dependent on his natural environment—he gets food from soil, fruits and timber from trees, medicines from plants, meat from birds and animals, fish from ponds, rivers and seas, water for daily use from springs, rivers and ground water.

Indian Tradition: India has a rich tradition in environmental ethics. Ancient Indians used to worship Nature—sun, water, air, soil, plants and animals. The ancient philosophers believed that the entire universe including earth, plants, animals and human beings is made of five essential elements viz. *kshiti* (earth), *op* (water), *teja* (fire or energy), *marut* (air) and *byom* (open space). Bodies of plants, animals and human beings are created out of these five essential elements and after death and decay, their bodies are disintegrated and converted back to these elements.

The fundamentals of conservation ethics were brilliantly formulated in the *Isho-Upanishad* during the Vedic era. *"The whole universe together with all its creatures belong to the Lord (Nature). Let no one species encroach over the rights and privileges of other species. Man can enjoy the bounties of Nature by giving up greed."* In other words, if we over-exploit Nature, we shall lose all the benefits from Nature.

In the Maurya period the emperors emphasised the awareness campaign for conservation of Nature and upholding the spirit of non-violence towards trees and animals. Compassion for animate and inanimate objects is part and parcel of Indian culture. But unfortunately we have forgotten this basic philosophy and turned to Western culture of conflict with Nature.

Scan of Human Civilisation: It is worthwhile to scan human civilisation since man's arrival on earth. Primitive man lived in mountain caves and looked at the forces of Nature with wonder and awe. He was panicky about the sun, lightning, thunder, rains and also wild animals in the forests. He was constantly busy in protecting himself and his family against the hostile elements around him. Gradually he discovered fire and made stone tools for his defence against wild animals. About 10,000 years ago agriculture was discovered and that was the starting point for human civilisation.

In course of time man devised science and technology step by step and depended on these for improving his living standard. This became prominent after the **Industrial Revolution (1760–1840)** which transformed England from an agricultural country into a predominantly industrial one. The invention of steam engine by James Watt (1785), mechanisation in textile industry, advancement in mining, transport and ship-building industries made Britain the most prosperous country in the world. This was promoted by British colonial exploitation.

With the help of science and technology, man overcame the natural barriers and established his supremacy over Nature. He can travel from one part of the world to the other part, in a matter of a few hours, undertake journey to outer space and planets, dive down to the bottom of seas and explore the wonders there and so on. But at the same time, the environment became more and more degraded and polluted. As a result of man-made activities (deforestation, urbanisation, industrialisation, etc.) the quality of environment suffered which has threatened the survival of man himself on earth.

1.4 IMPACT OF TECHNOLOGY ON ENVIRONMENT

Human Civilisation flourished with the help of science and technology. But technology has some adverse impact on environment. Industrial emissions and effluents, vehicular emissions and mining activities pollute the environment.

Industries produce environmental hazards on soil, water bodies and in air. They consume 37% of world's energy and emit 50% of world's CO_2, 90% of SO_2 and all the chemicals threatening O_3 depletion (O_3 *hole*). Every year they generate 2100 million tons of solid waste and 356 million tons of hazardous waste.

In developing countries, such as India, most of the industries discharge untreated waste on ground and into water bodies and also air.

There is worldwide concern about the disposal of radioactive wastes from nuclear reactors. Nuclear accidents are of frequent occurrence in recent years. The well-known example is Chernobyl Disaster (1986) in USSR which killed about 2000 people and contaminated the region with radiation.

Transport is a great consumer of land and energy. Transport consumes 30% of world's energy (82% consumed on roads) and produces 60% CO-emissions, 42% of NO_x and 40% for hydrocarbon emissions. Cleaner transport has been introduced in some countries such as Brazil where cars are run on biofuel, i.e., ethanol or ethanol/petrol mixture obtained from specially grown crops. Several countries such as Italy are running cars with compressed natural gas (CNG) as fuel.

Mining activities are important part of a nation's economy. These involve extraction of minerals fossil from deep deposits in soil employing the techniques of subsurface mining and surface mining. But these are associated with environmental hazards:

(*a*) *Devegetation and Defacing of Landscape:* Large scale devegetation or deforestation leads to ecological imbalances besides disfiguring the landscape. Moreover, the huge debris and tailing spoil the environment of the region and cause soil erosion.

(*b*) *Subsidence of Land:* Underground mining (e.g., coal) causes subsidence of land above resulting in tilting of buildings, cracks in road and also bending of rail tracks, etc.

(*c*) *Ground Water Contamination:* Mining disturbs the hydrological processes and also pollutes the ground water. Acid mine drainage (see Chap. 4) from abandoned mines contaminates nearby streams and lakes and damages aquatic life (plants and fish).

(*d*) *Air Pollution*: In metallurgical industries smelters emit large volumes of sulphur oxides, soot, As, Pb, Cd particles, etc. which cause public health hazards.

(*e*) *Occupational Health Hazards:* Miners are vulnerable to respiratory and skin diseases due to constant exposure to suspended particulate matter which is toxic. Hence they suffer from asthma, bronchitis, asbestosis, silicosis, etc.

1.5 ENVIRONMENTAL DEGRADATION

During space flight to the moon on board Apollo 14 in 1971, Edgar Mitchell caught a glimpse of the earth which filled him with ecstasy. He sent a message to the NASA ground station at Houston, USA: "It looks like a sparkling blue and white jewel laced with swirling veils of white—like a small pearl in a thick sea of mystery".

Three decades later if Mitchell was sent back to the space but this time equipped with special spectacles, to see invisible gases of the earths atmosphere, his feeling would have been quite different. He would observe giant punctures in the ozone layer over Antarctica and N. America. Instead of "Swirling blue and white jewel" he would see a dull dirty Earth with dark swirling clouds of CO_2 and SO_2. He would have radioed back to Houston *"What on Earth have we done during 30 years to change this scenario?"*

But we need not undertake space flight to see what we have done on Earth. Today we eat, drink, breathe and smell pollution. We have degraded land, destroyed forests at suicidal rates, thrown tons of toxic wastes into rivers indiscriminately and poured toxic chemicals into the seas. Moreover, we discharged greenhouse gases into the atmosphere for more than a century triggering climate changes. We also made holes in the ozone layer with gases from the refrigerators and airconditioners, thereby exposing us to skin cancer.

Environmental Education in India

The Bhopal Disaster (1984), the worst industrial disaster in history, which killed 10,000 people and injured lakhs of people, sounded a wake up call for India to take note of the crisis and realize the need of environmental awareness and education at all levels of the society.

In spite of its urgent need, the study of Environmental Education remains largely neglected. Only 25 out of 250 universities offer M.Sc. courses in Environmental Sciences. In the context of the environmental crises all over the country, the Hon'ble Supreme Court of India issued in 1991 directive to the Central Government for introduction of common course on Environment at the undergraduate level for various streams (Arts, Science, Commerce, etc.). But the universities were slow to respond. In absence of proper background of the subject and suitable textbooks, the teachers and students presented a scenario of confusion in colleges and universities. Once again after a gap of 12 years, displeased with non-compliance of their earlier order, the Supreme Court issued fresh order in 2003 to all states and Educational Institutes (Universities, IITs, Technical Universities) to introduce the course on Environment from higher secondary level to under graduate levels, from 2004–2005 session.

1.6 SOME ENVIRONMENTAL ASPECTS

Toxic Elements—Some elements are discharged by industries into air, water and soil. They get into the food chain and enter our biological system causing health hazards, in some cases, leading to death, Examples are Cr, Pb, Cd, Hg, Zn, As. etc.

Particulates—Particles are important contents of the troposphere. They vary in number from several hundred per cubic centimetre in highly polluted air. There size is in the range of 0.1–10μ.

They are responsible for electrical, chemical and photochemical reactions in the atmosphere, cloud and fog formation and maintain in heat balance of the atmosphere.

Air Pollution—The presence of gases such as CO, SO_2, H_2S, hydrocarbons in the atmosphere, introduced by external sources—natural and industrial (man-made). Natural sources include volcanic activity, vegetation decay and forest fires. These air pollutants disturb the dynamic balance in the atmosphere and affect man and his environment. This is the phenomenon of air pollution. (see Chap. 4: Air pollution and control for details).

Acid Rain—NO_x (nitrogen oxide) and So_x (sulphur oxide) emissions from industries and vehicles give rise to HNO_3 and H_2SO_4 in the atmosphere by chemical reactions. These together with HCl emission (from man-made and natural sources) generate acid rain which damages the environment. (see Acid Rain, Chap. 4 for details).

Water Pollution—The normal properties and function of pure water and dynamic balance in its system are disturbed by the presence of pollutants from external sources-man made and natural giving rise to the phenomenon of water pollution. Water pollutants are of different categories, organic (oil, grease, sewage, etc.), inorganic (mineral acids, metals, etc.), sediments thermal, etc. (scc Chap. 5 for details).

Noise pollution—Noise strikes human ear drum. At high level, it causes pathological and psychological disorder. This is known as Noise pollution. (see Chap. 7 for details).

1.7 ENVIRONMENTAL IMPACT ASSESSMENT (EIA)

The inherent conflict between development and environment can be solved to a great extent by a sound environmental management plan which is based on balancing development with environment. The necessary tool for this is *Environmental Impact Assessment* (EIA). Development should not be treated as an economic goal but as a multi-dimensional concept covering economic as well as political social and cultural aspects of life of common man.

The basic objective of EIA is to identify, predict and evaluate the probable economic, environmental and social impacts of developmental activities and take necessary steps as remedial measures which will be a part of the overall environmental management plan (EMP).

It is the government's policy that any industrial project particularly major industry must obtain EIA clearance from the ministry of environment before approval by the planning commission. It may be mentioned that EIA is conduted by a team of experts in the field (environment), appointed by the Ministry of Environment, Government of India.

QUESTIONS

1. What is meant by environment? What are its functions in relation to man?
2. What are the components of environment?
3. Discuss the relationship among the different components of environment.
4. Give an account of the impact of human activities on environment.
5. How is environment affected by technology?
6. Name the major regions of the atmosphere with their respective attitude and temperature ranges.
7. Write notes on :
   (a) Biosphere
   (b) Hydrological cycle
   (c) Environmental impact assessment

❍ ❍ ❍

2 Population and Natural Resources

> **"There is enough on Earth for everyone's need but not for everyone's greed."**
>
> —MAHATMA GANDHI

2.1 HUMAN POPULATION AND DISTRIBUTION

Population is intimately related to environment. The human population has grown faster in the 20th century than ever before. World population doubled in 40 years between 1950 and 1990 to cross 5 billion. The developed countries account for 1.5 billion while developing countries 3.5 billion population. By 2000 A.D. the population has touched 6.5 billion and by 2010 it will grow to 7 billion (one in every seven will be an Indian). World population is growing by 92 million every year, roughly adding population of Mexico.

It is interesting to note that it took about 2 million years for the world population to become 1 billion (1830), 100 years for 2 billion (1930), 30 years for 3 billion (1960), 25 years for 4 billion (1985) and 12 years for 5 billion (1997). The population stands at 6.3 billion (2000) and is estimated to be about four times around 22.5 billion in 2100. In developed countries the population is likely to be less than double while in developing countries like India, about four times (2100).

The statistics for India is of serious concern. Between 1901 and 1951 India's population grew from 238 million to 361 million, an increase of 52 per cent in 50 years. Between 1951 and 1981 it expanded from 361 to 685 million. Post-Independent India in 35 years (1947–1981) literally added a second India, i.e., doubled its population. At present, it has exceeded 1 billion mark (May 2000), close to China. *Present population 1.13 billion (May 2008). We have been overwhelmed by population explosion since 1980.

---

* World population 7 billion in 2012 and 9 billion in 2050—population in developing countries rising from 5.6 billion (2009) to 7.9 billion (2050) [UN official estimate].

Distribution

For historical and other reasons, world population is not uniformly distributed. USA and Canada have a population of 250 million; South America and the Soviet Union (CIS) have the same population. Africa and Western Europe have about 500 million people; East Asia, i.e., China, Japan and Korea have more than 1 billion while South Asia is the most populous region with 1.5 billion population (India, Pakistan, Bangladesh). India is adding every year the population of Australia (16 million).

2.2 POPULATION DENSITY

The density of population within a country gives a rough idea of the effect of population on natural resources. In general, lightly populated countries retain more of their original vegetation and wild animal populations than those with dense human population. Thus Australia (4 people/sq. km) is not likely to face wood or water shortage compared to Bangladesh (1800 people/sq. km). *Population density* is defined as *the number of persons per sq. km.* It should be noted that the total population size or population density cannot accurately predict the impact of environmental damage on the economy of a country. Some of the most crowded countries in East and South Asia are the most prosperous, e.g., Taiwan, South Korea, Hong Kong and Singapore where the population density is 1000 people/sq. km.

Some crowded countries can even manage to produce all their own food. Thus U.K. (600 people/sq. km) exports food while Brazil (44 people/sq. km) has acute shortage of food and has to import it. Again, Europe has all the timber and fresh water it needs but Africa (95 people/sq. km) has acute shortage of both these items. Such inequalities among countries are due to difference in their economic status—sustainable and non-sustainable. The countries with stable population (e.g., U.K.) enjoy sustainable economy and are prosperous. But countries with fast population growth suffer from non-sustainable economy and consequent problems of poverty, diseases, etc.

2.3 AGE STRUCTURE

In contrast to the developed countries, India has a pyramidal distribution of age-wise population. Children constitute 40 per cent, youth 33 per cent, middle aged persons 21.5 per cent and old 6.2 per cent. Such a structure with predominantly young people up to 34 years favours fast population growth. While India has a youthful age structure (reproductive group), in the Western countries the population has much less youthful age structure which slows the population growth.

The ratio of people over 65 and under 15 to the rest of the population is known as *dependency ratio*. This gives a measure of the economic impact of the age structure. People above 65 and below 15 contribute little to the economy and must be supported by the working population. Dependency ratios are rising in the developing countries. High dependency ratios as in India have adverse effect on the economy. This explains why working Indians have lower standard of living compared to developed countries.

Fertility Rate

The general fertility rate of a population is the number of babies born to 1000 women of the reproductive age. The rate of population growth, however, is based on the average age of reproduction. Thus whether a population is increasing, decreasing or stationary is more correctly estimated from the age-specific fertility rate. The fertility rate for India and other developing countries is about 3.0 while that for the Western countries is below 1.5.

Doubling Time

The period within which the population of a country doubles is known as its *doubling time*. This gives an idea of the population growth rate. Most developed countries have doubling times of more than 100 years while in the developing countries as in India, the doubling time is less than 25 years. It is absurd to double our resources viz., water, food and energy in 25 years and hence we fail to tackle the pressure on schools, hospitals, police protection and other vital services.

Infant Mortality/Life Expectancy

The death rate is generally expressed as the number of deaths per year for every 100 people. Modern medical science has helped to reduce the death rate but not the birth rate. Death control without birth control has led to population explosion.

Infant mortality is one of the most tragic indicators of poverty of a country. It also sets the parameter of *life expectancy*, i.e., the average number of years that a new-born baby is expected to survive. Progress in medical science has in general increased the life expectancy. In 1900 life expectancy was 46 years in the developed countries because the infant mortality was high, i.e., 40 per 1000. The situation improved much in 1984 in these countries when the life expectancy rose to more than 70 years as the infant mortality dropped to about 10 per 1000. The world's highest life expectancy is that of Japan (79 years) which enjoys high literacy and high standard of living. In the developing countries high infant mortality is responsible for low life expectancy. Furthermore, lack of basic amenities of life viz., access to clean water, nutritious food and elementary education explains why life expectancy in most of Africa is 45 years while in East Africa and South America life expectancy is 65 years.

It is on record that in developing countries 1 in 3 children is malnourished and about 3 million children die annually from diseases that could be avoided by immunisation.

And moreover, 1 million women die each year from preventable reproductive health problems.

2.4 CARRYING CAPACITY

The maximum population size that an ecosystem can support under particular environmental conditions is known as the *carrying capacity*. In natural ecosystem with unlimited resources and ideal environment, species can multiply at a maximum rate. However, in actual practice, the population of a species remains in check due to interaction of the inhabiting species as also finite nature of resources.

It is an established fact that while foodgrain production can under optimum conditions increase arithmetically (1, 2, 3, 4,...), population increases geometrically (2^2, 3^2, 4^2,...).

$$\text{Rate of population growth, } \frac{dN}{dt} = r\,N$$

where N = population size

r = specific growth rate

With doubling of population, resources do not double and hence set in a critical situation. If the population size far exceeds the carrying capacity by a wide margin, it leads to *population crash* or *explosion* when environmental conditions get degraded lower the carrying capacity.

The earth weighs about 5.97×10^{24} kg. If the present population growth continues for the next 5000 years, then the weight of human population itself will match the earth's weight. In other words, another earth will have to be accommodated within this earth, which is totally absurd. This implies man will run out of space and resources. Thus the present population growth will get more and more unfavourable for human survival as it will destroy the carrying capacity in the long run.

The concept of carrying capacity, as stated above, implies that only a limited number of people can be supported by the resource base of a country. The population growth must be linked to the resource base in order to have sustainable development. The developed countries follow the policy of *Population Stabilisation*, i.e., keep the population growth level below 0.5 per cent and thereby hold the key to economic development and resource management. This is the success story of all developed countries.

For India, population explosion is a time bomb, which must be diffused sooner the better. It is the major crisis facing the country. *We have to break the vicious P-triangle: Population-Poverty-Pollution on an emergency basis for our own survival.*

2.5 POPULATION GROWTH PATTERN (INDIAN CONTEXT)

During the British rule in India the population growth rate was insignificant. Agriculture was neglected and peasants were tortured as a matter of the colonial policy which led to poverty, malnutrition, famine, epidemics and death. The birth rate and death rate were almost balanced and this caused population drop by 1.1 per cent during 1911–1921. Slow increase in population was noticed from 1931 onwards which got a boost after Independence in 1947.

The population of 340 million in 1947 more than doubled to 683 million in 1981 within a period of 34 years and to over 1 billion in 2000. In other words, India added a second India within 34 years and a third India within 53 years since Independence. The population increased by 160.6 million during 1981–91 and after touching the 1-billion mark by May 2000, India became the second most populous country in the world, next to China. At present, India has been adding a population of 18 million every year roughly equal to the population of Australia. At this rate it is expected to overtake China by 2025 A.D. and earn the distinction of becoming the most populous country in the world.

The population growth rate in India is shown in Table 2.1.

Table 2.1: Population Growth in India (1901–2000)

Year	*Population (in million)*	*Decade growth rate (%)*
1901	238.1	–
1911	252.1	+ 5.75
1921	251.3	– 0.31
1931	279.0	+ 11.00
1941	318.9	+ 14.22
1951	361.1	+ 13.31
1961	439.2	+ 21.51
1971	548.2	+ 24.80
1981	683.8	+ 24.75
1991	846.3	+ 23.51
2000	1000 (1 billion)	
2009	1200 (1.2 billion)	

Such fast growth rate of population (2.4%) has led to population explosion. This is responsible for rapid environmental degradation and poor economy of the country.

2.6 URBANISATION

Industrial progress along with job opportunities led to growth and mushrooming of cities all over the world as centres of commerce and production and of art and culture. These also encouraged migration of rural population to cities. At the time of Independence the share of urban population did not exceed 8 per cent of total population but by 1991 it rose to 36.2 per cent (217 million). The number of million-plus cities increased from 3 (pre-1947) to 23 by 1991. The top ten such cities are listed in Table 2.2.

Unplanned growth of urbanisation has led to collapse of infrastructure particularly in the metropolitan cities and also miserable living conditions in slums with the net result that the citizens have poor quality of life.

Table 2.2: Top Ten Cities (1-Million-plus) of India (1991)

Cities	*Population (in million)*
1. Greater Mumbai	12.57
2. Kolkata	10.92
3. Delhi	8.38
4. Chennai	5.36
5. Hyderabad	4.28
6. Bangalore	4.09
7. Ahmedabad	3.30
8. Pune	2.49
9. Kanpur	2.11
10. Nagpur	1.66

The world scenario of megacities (population exceeding 10 million) is presented in Table 2.3.

Table 2.3: World's Megacities (2000)

Cities	*Population (in million)*
Tokyo	28.0
Sao Paulo	22.6
New York	16.6
Mexico	16.2
Sanghai	17.4
Mumbai	14.2
Los Angeles	13.2
Buenos Aires	13.8
Seoul	13.0
Beijing	14.4
Kolkata	12.7

It is intriguing that whereas London, once a megacity till 1950s, moved out of the list with its stabilised population, Kolkata moved in since the 1980s. Originally built for a population of 1 million, the infrastructure of Kolkata is on the point of collapsing under the pressure of the megacity population for their basic civic amenities viz., housing, transport, roads, water supply, sanitation, etc.

2.7 IMPACT OF POPULATION PRESSURE ON INDIA'S ENVIRONMENT—FOOD AND NATURAL RESOURCES

India occupies 2.4% of world's land area but it accounts for 16% of world population. Also the hard fact remains that among 1 billion plus population, about 400 million are illiterate and an equal number live below poverty line. Such huge population has enormous demands for food, shelter, water, fuel, sanitation, etc. which in turn have impact on the natural resources.

Food: Rice is the principal crop which mainly grows in the monsoon-dominated regions of West Bengal, Assam, Bihar, Orissa, eastern U.P. and southern states of Andhra Pradesh, Tamil Nadu, Kerala and Karnataka. The second and most important crop is wheat which grows in Punjab, Haryana, U.P., M.P. and in the plains of Gujarat and Rajasthan. The Green Revolution of the sixties boosted wheat production in Punjab and Haryana, which are known as the bread baskets of India. Use of high input, high output technology viz., high yielding variety of wheat (Mexican), intensive irrigation, chemical fertilisers and pesticides showed record production of wheat in Punjab and Haryana. This was reflected in increased overall food production. From 50 million tons in 1950, the foodgrain production jumped to 100 million tons in 1970, 180 million tons in 1993 and 200 million tons in 1998. But since the nineties, they reached the plateau. The foodgrain production cannot keep pace with our population explosion. It is well known that while agricultural production increases arithmetically, population increases geometrically. The cultivated land (about 143 million hectares) has already reached the saturation carrying capacity since it is subjected to soil erosion, loss of soil fertility due to intensive cultivation, excessive use of fertilisers and pesticides.

Water: The water front presents another gloomy scenario. Heavy withdrawal of ground water for intensive cultivation, for more than two decades, caused water tables to sink thereby threatening food production and food scarcity of the country. According to the estimate of the International Water Management Institute (IWMI) withdrawals of ground water are double the rate of aquifer recharge from rainfall. As a result, water lables all over the country are falling. This is likely to reduce India's grain harvest by one fourth.

On the other hand, population explosion has, in general, exerted tremendous pressure on water resources. Our available water resource is 690 cubic km (km^3) surface water and 452 km^3 ground water. This is not adequate to meet the growing demand for water and is likely to result in water scarcity in near future all over the country.

Soil: Fast growth of human and livestock population has led to continuous degradation of soil due to human activities such as intensive cultivation, over utilisation of grazing land, reckless deforestation, fuel and fodder collection, road construction, surface mining, etc. It is estimated that every year we are losing 6000 tons of top soil resulting in soil erosion, loss in soil fertility, drought and flood.

Forest: Population explosion exerts tremendous pressure on forest resources to meet the ever-increasing needs for land, timber, fuel, etc. This has resulted in random deforestation with annual loss of 1.3 million hectares of forests. Deforestation brings in its trail extinction of valuable species of plants including medicinal plants, animals causing loss of biodiversity and hence national calamity.

Fuel: Population pressure has imposed severe strain on non-renewable and conventional energy resources like fossil fuel (coal, petroleum, natural gas) and also mineral resources such as iron, copper, lead, silver and gold. The increasing consumption of fossil fuel for domestic and industrial needs and transport has generated air pollution and health hazards particularly in cities and industrial areas.

It may be concluded that population, food, natural resources and environment are closely interrelated. As a matter of policy, the population growth has to be reduced from 2.4 per cent to 1 per cent (China, 1 per cent; western countries, 0.5 per cent), in order to protect the environment and save the next generation. We have to pull ourselves out of the vicious P-Triangle in which we are trapped: *Population-Poverty-Pollution.*

2.8 WOMEN AND ENVIRONMENT

Women, particularly poor village women, are the worst victims of environmental degradation. Their day starts with a *long march* in search for fuel, fodder and water. Many such women spend up to 10 hours a day fetching fuel, fodder and water. An average family in Karnataka village walks about 5 km daily on average for collecting fuelwood only, equivalent to walking from Delhi to Mumbai every year. In deforested hill and forests the treks are longer and more difficult. By lighting firewood in chullahs, they turn their kitchens into gas chambers where they spend 2–3 hours daily for cooking food. They are exposed to heavy fumes and smoke and suffer as much as a chain smoker smoking 20 packets of cigarettes a day.

Biomass plays a crucial role in meeting daily needs of the vast majority of rural households. The different forms of biomass are food, fish, fuel (firewood, crop wastes and cowdung); fodder, fertiliser for agricultural fields (cowdung, organic manure, forest wastes); building materials (timber and thatch) and medicines (herbs). The country's biomass base is affected by deforestation and loss of vegetation cover. It is the village women who are worst hit by deforestation and devegetation (loss of vegetation cover). Similarly, the forest dwellers, viz., tribals and nomads those who wander from one place to another and have no fixed location, face total social destruction—they are converted into the uprooted lot of *"environmental refugees."*

Woman Power

In the 18th century some 350 Bishnoi men and women from the **Bishnoi** community in Rajasthan were determined to protect their forest. They clasped the trees when the

Maharaja of Marwar sent some woodcutters along with soldiers to cut down those trees which were required for a brick kiln (furnace) to make slabs for his palace. But the group of men and women resisted by continuing to clasp the trees and finally they were hacked to death. This was the precursor (forerunner) of the Chipko Movement two centuries later (1983).

The women of the Garhwal hill region of Uttarakhand were at the vanguard of the **Chipko Movement**. They organised and executed the movement with utmost sincerity and spread their base all over the Garhwal region. Thus organised woman power prevented further ecological crisis in the Himalayan region.

Another shining example of woman power was the case of Khirakot women in Almora district of Uttarakhand Himalayas. A Kanpur contractor was setting up soapstone mines in the hills thereby destroying the ecology of the region. The Khirakot women sprang into action—they got organised, stopped the work of the mines and also fought legal case against the contractor. Finally they won the battle, both legal and environmental and got the contractor's license cancelled.

2.9 NATURAL RESOURCES—CLASSIFICATION

Resource is defined as *a form of matter and energy which is essential for the functioning of organisms, populations and ecosystems*. Man depends on the resources for his day-to-day living. These resources are air, water, soil, forests, vegetables, milk, fish, animals, energy (fossil fuels, solar power, etc.) etc.

Man himself constitutes an important resource viz., human resource. Human resource development through education, skill and knowledge has now been given top priority in national policy since it is an index of the development status of a country. This has been emphasized by the United Nations Development Programme (UNDP).

Classification

The resources are classified as Renewable and Non-renewable or Conventional and Non-conventional.

Renewable Resources

These are exhausted within a limited period but can be regenerated and used again, i.e., recycled. Examples are: forests, plants, water, solar energy, etc. Forests are renewable resources since after felling of trees for wood, fuel, etc. They can be recovered by fresh plantation (afforestation).

Non-renewable Resources

These cannot be renewed after being exhausted. Examples are: fossil fuel (coal, petroleum), natural gas and mineral. All these took thousands of years for their formation under the earth's surface. For example 100-million year old coal deposits can be burnt to carbon dioxide in a matter of a few seconds by power plants. But by no means coal can be restored or renewed. Non-renewable resources get exhausted due to over-exploitation by man and then they can neither be regenerated or renewed.

2.10 NATURAL RESOURCES—USES AND ABUSES

Mineral Resources

Mining and processing of minerals involve environmental problems such as disturbance of land, air pollution from dust and smelter emission as well as water pollution from disrupted aquifers.

The rate of depletion of mineral resources is measured by two parameters—per capita mining and per capita consumption.

Per capita mining is calculated by dividing the amount of resource mined by the population. *Per capita consumption* is obtained by dividing the resource amount by the population.

It is known from the world per capita mining figures that five minerals are mined to the maximum extent—coal, petroleum, iron ore, aluminium and phosphate rock. The demand on natural resources varies from Asian to American continent. USA imports substantial quantities of mineral resources so that its per capita consumption exceeds its per capita mining figure.

The major non-metal resources include asbestos, carbonates, chlorine, granite, oxygen, phosphate, sand, etc. Asbestos, sand, gravel and calcium carbonate together with granite constitute the common and most widely used building materials. Chlorine, magnesium, sodium chloride are the major representatives of the group of resources with a vast reserve viz., the oceans.

Land Resources

Land is used for many purposes—construction of houses and buildings, roads, railway lines, airports, agriculture, grazing of cattle, setting up of industries, etc.

In India the total available land is about 300 million hectares. It is, however, a fixed asset. As a result of our population explosion, we possess the lowest man: land ratio, barely 0.50 hectare per capita due to use and abuse over the centuries.

The percentage break-up of land use pattern in our country is:

(i) Waste land 24%

(ii) Pasture land 4%

(iii) Forest land 21% and

(iv) Agricultural land 51%

Considering the size of India, available pasture land is insignificant while forest land has been reduced due to conversion of forests into agricultural land and also for housing, roads, industries, etc.

Soil Resources

We depend on soil for our daily food and also for meeting our needs for fibre, timber, firewood, etc. Hence, soil is indeed an important resource. But this is getting depleted mostly due to erosion and degradation. Large scale deforestation and intensive grazing of grassland have exposed the soil to the weathering forces—water and wind. Apart from these, the prevailing practice of jhum cultivation in the N.E. region, canal irrigation, road construction and surface mining also favour soil erosion. Top soil (most fertile zone) is lost as a result of soil erosion leading to loss of soil productivity. It is estimated that every year 25 billion tons of top soil are lost from farm land by erosion. In India about 80 million hectares of cultivated land is subjected to severe erosion and about 30 per cent of the total eroded soil is permanently lost to the sea. This is critical issue as it affects our survival in terms of supply of food, fodder, fuel and daily necessities.

Wildlife Resources

The bounties of Nature include wonderful variety of flora and fauna with their habitats in different types of forests but deforestation has destroyed their habitats in the past whereby many plants and animals are on the way to extinction.

Wildlife population has been considerably depleted because of reckless hunting of animals and birds for food, furs, feathers, skin, tusks, etc. since the 1950s. Thus out of 65,000 species of animals, almost 600 have been identified as endangered species. We also have lost many species of colourful birds out of a total number of 1200. This wave of man-made extinction has already upset the ecological balance in the environment and badly damaged India's economy.

Fish Resources

Fresh and marine waters provide fish, an important renewable resource. It is a well-known food item with rich protein content. Fish oil is a useful tonic and fish residues are used as fertilisers. People in developing countries cannot afford meat for their diet and so they depend on fishes the only source of animal protein in their diet. In the earlier days fish was used on a small scale but now with population explosion, it has been transformed into a large scale industry. There are abundant fish resource in the country with about 30,000 species of fish.

But in recent years fish resources, as a whole, have been hit by increasing water pollution. This has caused fish scarcity in general.

Over-consumption and overfishing have turned fish like halibut and salmon into endangered species. World-wise annual catch of fish is about 8.5 crore metric tons. The major fishing grounds are the North Coast of USA (the Grand Banks) and North Sea (Dogger Banks)—these centres conduct major trade in fish products. Japan and Norway are the principal fishing nations in the world. Japan is in the forefront accounting for 26.5 per cent of the total global catch. In India fishing is done in inland water—rivers, lakes, ponds, etc. to satisfy the requirements for local consumption. Marine fishing is practised in coastal areas but is handicapped due to shortage of modern equipments like trawlers and large ships and also lack of training and facilities.

Recently the government is paying attention to the development of pisciculture or fish culture by setting up fisheries in lakes, ponds and rivers. It is necessary to control overfishing and water pollution.

Agricultural Resources

From ancient times India has been predominantly an agricultural country. Hence agriculture is the pillar of Indian economy. Agricultural resources are man-made using land, soil and water as the basic resources. Production of a variety of crops provide us with our staple food and fruits. Food crops are wheat, paddy, maize and barley. Fruit trees yield mango, banana, coconut, orange, apple, grape, etc. In addition to these there are plantation crops such as tea, coffee and rubber; fibre crops such as cotton, jute and commercial crops such as tobacco, sugarcane, sugar beet, etc. Each crop needs specific type of soil, climate and environment for its cultivation.

At present agriculture in India involves about 70 per cent of people and 47 per cent of land area of the country (c.f. USA twice the size of India and less than half the population engages 2.4 per cent of people and 20 per cent cropland). Our production of food grain has gone up from 50 metric tones (Mt) (1950–51) to 155 Mt in 1984–85 and 250 Mt in 2000 A.D. The production is about three times the soil capacity (75 Mt). This has been possible with the application of advance technology of high input-high output, i.e., use of high-yield variety of seed, increased amounts of fertilisers and pesticides and extensive irrigation resulting in **Green Revolution**. Punjab and Haryana are the leading beneficiaries of Green Revolution with record production of wheat.

However, this western model of high-input high-output agriculture, in spite of paying rich dividends in terms of production and making the country self-sufficient in food production, has brought in its trail a downward trend since the last decade. The latter is due to loss of fertility of soil, loss of top soil, soil degradation, depleting groundwater reserve as a result of heavy withdrawal for irrigation, increased surface water pollution, etc.

The present crisis facing the agriculture front can be overcome with the help of advanced techniques e.g., organic farming, biotechnology, less use of chemical fertilisers and pesticides, etc. As a matter of policy, deforestation should be minimized, afforestation should be enhanced, population time bomb should be diffused while over-consumption of food and agricultural crops should be kept under control. This will help in bridging the gap between agricultural yield and its potential yield and demand for food by increasing population.

Animal Resources

Domesticated animals providing us with meat and milk products (animal protein) are known as animal resources. They supplement our food from agriculture. They also supply us with out requirements for wool, hides and skin.

In India animal breeding and rearing have not received much attention and have not been commercially developed. Animals are used to supplement human labour for agriculture. For example, cattle and buffaloes are used for ploughing, drawing water from wells and pulling carts. This is an unfortunate abuse of animal resource. India happens to have the largest cattle population in the world but due to negligence and misuse, milk yield is poor on an average and is not available to the poor sections of the population. The net result is that poor children are denied milk supply and suffer from malnutrition.

However, in urban centres dairy units have been set up to supply dairy milk, butter, cheese and milk powder to the urban population from their commercial centres in Gujarat and Maharashtra. Networks of Mother Dairy and Metro Dairy products have dominated the market (urban).

Water Resources

"*One can live several weeks without food, several days without water but only a few minutes without air.*"

Like air, water is an essential constituent of the life-support system. India is blessed with good rainfall, about 200 cm per year. The quantum of water is about 6 trillion m^3 which is among the largest in the world for a country of comparable size. But due to wastage and inefficient water management along with unequal distribution of rainfall, a large part of the country suffers from water scarcity while during the dry months even place like Cherrapunji in Meghalaya with heaviest rainfall in the world (1200 cm) experiences water shortage. Again during rainy season surface water from rivers runs off rapidly causing floods in Assam, W. Bengal and Bihar.

Among the major consumers of water, is the agricultural sector. Table 10.1 shows that irrigation (inclusive of livestock) and power use 79.6 per cent and 13.7 per cent water respectively, followed by domestic (3.6 per cent) and industrial (3.2 per cent) consumption.

With increasing population, the demands of water from each sector will also increase further (at least by 25 per cent by 2025 A.D.). Moreover, surface water pollution is likely to increase within the same period. These will put severe constraint on water resource in near future. At present 480 million of India's 1 billion people do not get safe drinking water. Only a sound Water Resource Management Policy can save India in the foreseeable future.

Table 2.4: Water Consumption in India (2000 A.D.)
(Available water is 1900 million m^3 per year)

Uses	*Input*	*Consumed (in 100 million m^3/year)*	*Unutilised*
Irrigation and Livestock	869	783	86
Power	150	5	145
Industry	35	10	25
Domestic	38	8	30
Total	1092	806	286

Forest Resources

Forests are valuable renewable resource for a country. They are closely related to vegetation, grassland and wild life resources. They have contributed substantially to the economic development of the country by providing goods and services to the people and industries. Furthermore, they play a very major role in enhancing the quality of environment by influencing the life support system. Forests are also linked with our culture. As a matter of fact, Indian philosophy and religion grew up under the guidance of our *rishis* in the forests in total communion with Nature.

India has 64 million hectares of forest land out of a total geographical area of 329 million hectares. It is on record that the country had 80 per cent forest cover 5000 years ago which has now been reduced barely to 21 per cent. Per capita forest land is 0.1 hectare compared to the world average of 1 hectare. This shows the poorest record of India. We are losing 1.3 million hectares of forest cover every year and at this rate of deforestation we are likely to lose all the forest cover and face total disaster within the next 25 years.

During 1951–2000, i.e., almost 50 years after Independence, India witnessed maximum loss of forests, namely, 6.0 million hectares of forest area. This overconsumption has been due to population explosion (both human and livestock population) accompanied by rapid urbanisation and industrialisation as well as change of life style. There has been tremendous pressure on forest systems to meet ever-increasing demands on fuel or fire wood, timber and pulp wood for housing, furniture, agricultural implements, railway slippers, sports goods, ship and boat building, paper and pulp industries, etc. Annual consumption of fire wood in the country is about 170 million tons while 10–15 hectares of forest cover is destroyed each year to meet fuel requirements. The consumption of fire wood is 80 per cent in villages and 20 per cent in urban areas. There is a shortfall of about 195 million m^3 in the demand and supply of fuel wood. Similarly, the gap for timber between demand and supply is about 15 million m^3. Because of the importance of timber, India has started importing timber from other countries.

The use of wood and its requirements are tabulated on next page:

Table 2.5: Use of Wood

Region	Total wood consumption (billion m³)	Wood use %	
		Industrial	Fuel wood
World	3.2	46	54
Developing countries	1.8	18	82
Developed countries	1.4	84	16

Deforestation has not only brought about environmental disaster along with loss of economic benefits. The disaster includes increased carbon dioxide content in the atmosphere and hence increases Greenhouse Effect, soil erosion, drought and flood, loss of habitats of wild birds and animals and also loss of valuable species including medicinal plants.

Table 2.6: Requirement of Wood (in Metric Ton)

Unutilised	Requirement (1980)		Requirement (2000)		Addl. requirement	
	m³	% of total	m³	% of total	m³	% of increase
Fuel Wood	188.6	87.5	22.5	78	36.4	10
Timber	22.7	10.5	46.8	16	24	106
Pulb Wood	4.2	2.0	17.7	6	13.5	323

Source: FAO Year Book of Forests (1981).

As forests are renewable resources, priority should be given to afforestation, i.e., plantation of new trees to make up the loss due to deforestation. The regeneration cycle of forests can vary widely in different climatic regions from 15 to 80 years. Afforestation is intended to ease the present pressure on forest resources and sustain the environment. It is also essential to control utilisation and consumption of forest wealth.

2.11 RESOURCE/ENVIRONMENT MANAGEMENT

The problem of conservation of the environment is that the human population has been going through explosion and along with it consuming the planet's resources, both renewable and non-renewable. In the relentless march towards development, industrialistion and urbanisation, the earth's natural resources have been plundered which have finally threatened human survival on earth. The problems of air, water and terrestrial pollution have assumed alarming dimensions day by day. In the not too distant future, foreign tourists may be advised not to drink local water nor breathe air while travelling abroad.

The Government of India in its Sixth Five Year Plan (1980–85) framework, laid down the following policy:

"It is imperative that we carefully utilise our renewable resources of soil, water, plant and animal life to sustain our economic development. Over-exploitation of these is reflected in soil erosion, siltation, floods and rapid destruction of our forest, floral and wild-life resources. The depletion of these resources often tends to be irreversible and since the bulk of the population depends on these natural resources to meet their basic needs, particularly of fuel, fodder and housing material, it has meant deterioration in their quality of life."

It is important to note the response of an unknown Red Indian Chief (USA) to the President of USA regarding the latter's proposal for purchase of land belonging to the Red Indians some 130 years ago. This is very much relevant even today.

"The Great Chief in Washington sends word that he wishes to buy our land. How can you buy or sell the sky, the warmth of the land? The idea is strange to us. If we do not own the freshness of the air or sparkle of the water, how can you buy them?...the land is sacred to us...the rivers are our brothers—they quench our thirst...**The earth does not belong to man, man belongs to the earth**...All things are connected. **Man did not weave the web of life, he is merely a strand in it. Whatever he does to the web, he does to himself**..."

2.12 SUSTAINABLE DEVELOPMENT

The Policymakers of the World Conservation Strategy (1980s) introduced the term "Sustainable Development". However, it was not commonly used till the Bruntland Commission (World Commission on Environment and Development, headed by Norway Prime Minister, Gro Hartlem Bruntland) published its Report "Our Common Future" in 1987. According to this Report, *Sustainable Development is development that meets the needs of the present without compromising the ability of the future generations to meet their own needs.* In other words, *sustainable development means a balanced development which satisfies the current needs with limited utilisation and conservation of natural resources without sacrificing the quality of environment and future of the next generations.* Such balance between environment and development only lays the foundation of Sustainable Development.

Sustainable development should be part and parcel of our national plan. While noting the population explosion trend and shrinking resource base, we should practice recycling and reusing materials including food and controlling pollution so that resources are not exhausted quickly and environment does not suffer. The idea of Sustainable Development was strongly supported the Earth Summit (1992) held at Rio de Janeiro, Brazil (United Nations Conference on Environment and Development i.e., CED). The Earth Summit adopted a historic Agenda 21–a comprehensive plan for leading to Sustainable Development (CED) busted with the task of moritoring progress in implementing Agenda 21.

The Commission has formulated the following guidelines for achieving sustainable development:

(i) Substantial reduction in population growth rates;
(ii) Housing, health care and education for the poor;
(iii) Education and empowerment of women, particularly in rural areas;
(iv) Afforestation (making up deforestation) in cities and rural areas;
(v) Use of non-conventional energy resources by industries and proper treatment of urban wastes and industrial effluents;
(vi) Conservation of land, soil, water, forest, wild life, fish and non-renewable resources by application of modern tools of science and technology;
(vii) Efficient use of resources avoiding wasteful and excessive utilisation;
(viii) A global and national strategy to eradicate poverty and
(ix) People's participation in conservation and improvement of environmental quality.

The Rio Summit through its Agenda 21 gave priority to population control, poverty eradication and people's participation involvement in the management of natural resources.

QUESTIONS

1. Give examples of Renewable and Non-renewable Resources.
2. Name the major metal and Non-metal Resources.
3. Give the percentage break-up of land use pattern in India.
4. What is Green Revolution? Discuss its benefits for agriculture. What is the recent fall out in food production?
5. What are the sources of fresh water? Give their quantities in terms of total percentages of water resources.
6. "Wood is a major renewable resource"—Elucidate.
7. Trace the world population growth up to 2000 A.D.
8. What is population explosion? How does it affect the economy of India?
9. What is doubling time of population? Explain with reference to world population and Indian population.
10. Write notes on:
    (a) Carrying capacity of the Earth
    (b) Population stabilisation and sustainable development
    (c) Woman power and environment.

❍ ❍ ❍

3 Ecology and Ecosystem

> **"If you plan for one year, plant rice; if you plan for ten years, plant trees; but if you plan for 100 years, educate the people."**
>
> —CHINESE PROVERB

3.1 INTRODUCTION

Ecology

The word '*Ecology*' owes its origin to the Greek words '*Oikos*' meaning 'House' and 'logos' meaning "Study or discourse". *Ecology is the branch of science that deals with the study of interactions between living organisms (microbes, bacteria, plants, animals and man) and their physical environment* (*air, water, land*). The latter are closely inter-related through continuous interaction by circulation of nutrients (C, O, N, P, S) so that any change in the environment has an effect on the living organisms and vice versa.

Ecosystem

The ecosystem is the *functional unit* in ecology as it consists of both the biotic community (living organisms) and abiotic (physical) environment. The latter have close interaction, essential for maintenance of life processes.

Components and functions—from biological point of view the ecosystem has the constituents:

(i) inorganic substances (carbon, nitrogen, carbon dioxide, water, etc.) involved in natural cycles;

(ii) organic compounds (proteins, carbohydrates, humic substances, etc.);

(iii) air, water and substrate environment including the climatic regime and other physical factors;

(iv) producers, autotrophic (i.e., self-sustaining organisms, main green plants that can manufacture food from simple inorganic substances);

(v) heterotrophic (i.e., depending on others for nourishment) organisms, mainly bacteria, fungi and animals which live on other organisms or particulate organic matter;

(vi) micro-consumers, decomposers, mainly, bacteria, fungi which obtain their energy by breaking down dead tissues or by absorbing dissolved organic matter, extracted from plants or other organisms. The decomposers release inorganic nutrients that are utilised by producers. They also supply food for macro-consumers or heterotrophic organisms bacteria, fungi and (animals) and often excrete hormone-like substances that inhibit or stimulate other biotic components of the ecosystem.

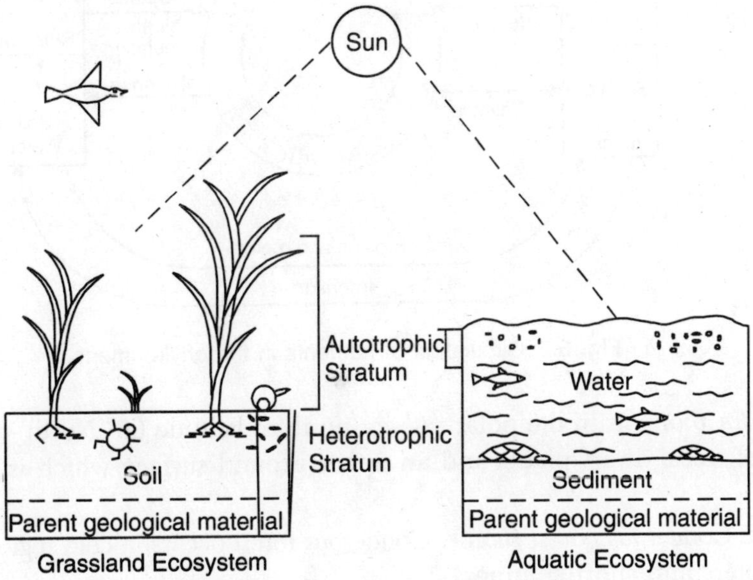

Fig. 3.1: Grassland and pond ecosystem

Typical profiles of a grassland ecosystem and of a pond ecosystem are shown in Fig. 3.1.

The common features of all ecosystems—terrestrial, freshwater, marine and agricultural—are the interactions of the autotrophic and heterotrophic components. The major autotrophic metabolism occurs in the upper "green belt" stratum where solar energy is available while the intense heterotrophic metabolism occurs in the lower "brown belt" where organic matter accumulates in soils and sediments.

3.2 ECOSYSTEM—BIOMES

The *Biome* is a very large land community unit where the plant species is more or less uniform. It provides a basis of natural ecological classification. The main Biomes of the world are the Tundra, Temperate, Coniferous forest, Deciduous forest, Temperate grassland, Tropical Savanna, Desert and Tropical Rain Forest.

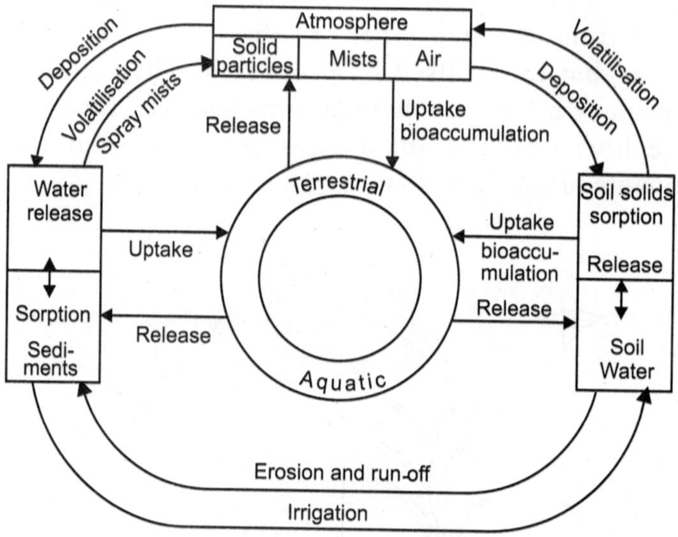

Fig. 3.2: Circulation of nutrients in the environment

The *Tundra Biome* is in the polar region (north of latitude 60° North)–it is characterised by absence of trees, dwarf plants and an upper ground surface which is wet, spongy and rough.

Temperate Coniferous Forest Biome: Coniferous forest occurs in cold regions with high rainfall, long winters and short summers.

Temperate Deciduous Forest Biome: These are high altitude regions about 3000–4000 metres above sea level as in the Himalayas. Here pines, fir and juniper trees are found.

Temperate Grassland Biome: This type of grassland occurs where there is about 25 to 75 cm of rainfall per year. Such grasslands are found as tall grass prairies, short grass prairies of North America and also in South America, steppes of Southern Russia and Asia.

Tropical Savanna Biome: These are tropical grasslands with scattered drought-resistant trees. These are found in eastern Africa, Australia and South America.

Desert Biome: These are found in very dry environment where temperature changes from very hot to very cold.

Tropical Rainforest Biome: These occur near the equator and offer the most diverse communities on earth with fairly high temperature and humidity. The annual rainfall is more than 200–225 cm. Here one finds dense vegetation consisting of tall trees covered with creepers and orchids, numerous herbs and shrubs. Tropical rainforest is the habitat of numerous vertebrate and invertebrate animals.

3.3 ECOSYSTEMS—CLASSIFICATION

Ecosystems of the world are studied on the basis of their principal habitats. Among the environmental segments, lithosphere and hydrosphere are the major habitats for a wide variety of flora and fauna.

Land-based Ecosystem

Land (terrestrial) ecosystems depend largely on the climate and soil. Higher plants and animals have evolved on land. For example, seed plants, insects, warm-blooded vertebrates and micro-organisms dominate on land now. The major terrestrial communities consist of herbaceous plants, shrubs, grass and also woody trees besides numerous insects, anthropods, birds, etc.

Marine Ecosystem

Oceans occupy 70 per cent of earth's surface, offering habitat to numerous plants, (mainly algae), animals like zooplankton, shrimps, oysters, fishes, reptiles, birds and mammals. They serve as the sink of a large quantity of run-off and wastes from land.

Marine water has a high salt content (about 3.5% by weight) and poor fertility due to lack of nitrates and phosphates as compared to fresh water. Marine life is abundant near the shore and in the continental shelf. The species include commercial fishes, large sea mammals like whales and seals.

Freshwater Ecosystem

Freshwater bodies (ponds, lakes, rivers, springs) are rich in nutrients (nitrates, phosphates) and provide good habitat for phytoplankton, zooplankton, aquatic plants and fishes.

Wetland Ecosystems

Wetlands are transitional lands between terrestrial and ecosystems where water stands at 2.5 to 300 cm during most of the year. They include valuable natural ecosystem harbouring a wide variety of plants, animals, fishes and micro-organisms. They are at present in danger due to increasing urbanisation as in case of eastern part of Kolkata.

Mangroves (Forest between Land and Sea)

Mangroves are important forest communities in tidal zones or equatorial and tropical coasts. For example, the Sunderbans in the Gangetic estuarine delta touching the Bay of Bengal offer important mangroves, habitat of wild animals including Royal Bengal Tiger and of interesting plant species.

3.4 ENERGY FLOW/ENERGY BUDGET

The movement of energy through ecosystem is called *energy flow*. It is an essential part of the ecological process of life support in nature. As stated earlier, green plants transform solar energy with the help of carbon dioxide and water into food energy (chemical energy) by photosynthesis. This food energy goes on transforming when food is taken as energy by other organisms of different trophic levels. Thus energy flow is related with Food Chain. Almost all the solar energy is trapped by autotrophic organisms of the first trophic level (green plants). Because much energy is dissipated, about 90 per cent of available chemical energy (food energy) is lost at each stage of transfer from one trophic level to the next higher level.

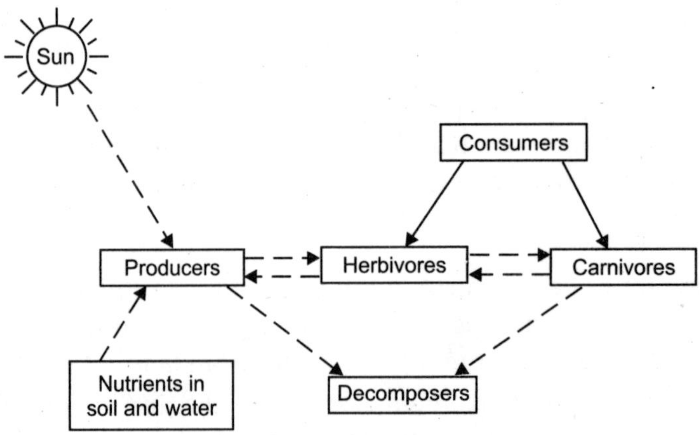

Fig. 3.3: Energy flow in ecosystem

Energy Budget

Energy Budget shows a record of the flow of energy through an ecosystem in each step from one organism to another.

The *primary productivity* of an ecosystem is defined as the rate at which radiant energy is converted by photosynthetic and chemosynthetic activity or producer organisms (green plants, for example) to organic substances.

In coastal waters of seas primary production in relation to biomass shows vertical distribution. It is concentrated in the upper 30 metres or so—in the clearer but less fertile waters of the open seas the primary production zone may extend down to 100 metres or more. This is why coastal water appears dark greenish and the ocean waters blue.

3.5 TROPHIC LEVELS

The transfer of food energy takes place in some hierarchical order in the ecosystem. The food energy passes from one group of organisms to other groups at different levels. These levels are called *Trophic levels* which mean feeding level of a group of organisms.

Trophic Level I: Here we find primary producers or autotrophs which produce their own food by converting solar energy into chemical energy in the form of organic substances such as carbohydrates. The green plants belong to this category.

Trophic Level II: Here we find primary producers who do not produce their own food but use plants as their food. Examples are grazing animals—cow, sheep, goat, deer, rabbit, etc. who are called herbivores.

Trophic Level III: At this level we find animals who depend upon other animals for their food. They are carnivores (tiger, lion, etc.) who eat mainly herbivores. This is the secondary consumer level.

Trophic Level IV: It is the tertiary trophic level to which belong all groups of animals including man who obtain their food from the other three levels. They are carnivores who eat both plants and animals.

It may be pointed out that organisms deriving their food energy from the same producer level belong to the same trophic level. Thus grasshopper, cattle and grain-eating birds are all primary consumers and occur at the second trophic level since they derive their energy from plants.

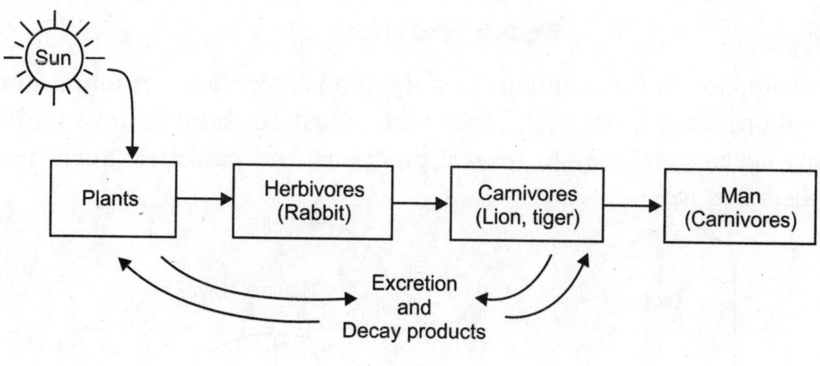

Fig. 3.4: Relationship among different biotic components of an ecosystem

3.6 FOOD CHAIN AND FOOD WEB

Food chain means the transfer of food energy from the source in plants through a series of organisms with the repeated process of eating and being eaten. In a grassland, grass is eaten by rabbits while rabbits in turn are eaten by foxes. This is an example of a simple food chain. This food chain implies the sequence in which food energy is transferred from producer to consumer or higher trophic level. It has been observed that at each level of transfer a large proportion, 80–90 per cent of the potential energy is lost as heat. Hence the number of steps or links in a sequence is restricted usually to four or five. The shorter the food chain or the nearer is the organism to the beginning of the chain, the available energy intake is greater.

Basically, food chains are of two types: (i) *Grazing food chain* and (ii) *Detritus food chain*. The grazing food chain starts from green plants and passes through herbivores to carnivores. On the other hand, the detritus food chain starts from dead organisms and goes into the micro-organisms and then to detritus feeding organisms and their predators. The term 'detritus' refers to all organic matter involved in the decomposition of dead organisms.

The energy from the sun is transferred in the form of food from one organism to another in the ecosystem.

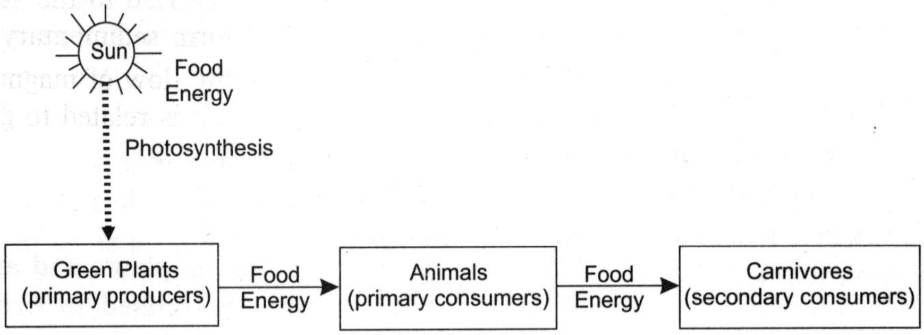

Fig. 3.5: Food chain

Food Web: In complex natural communities the food chains are not found in isolation. They are all interconnected with each other since most consumers have multiple food sources and many species are prey to several predators and parasites. Such interlocking food chain is called *food web*.

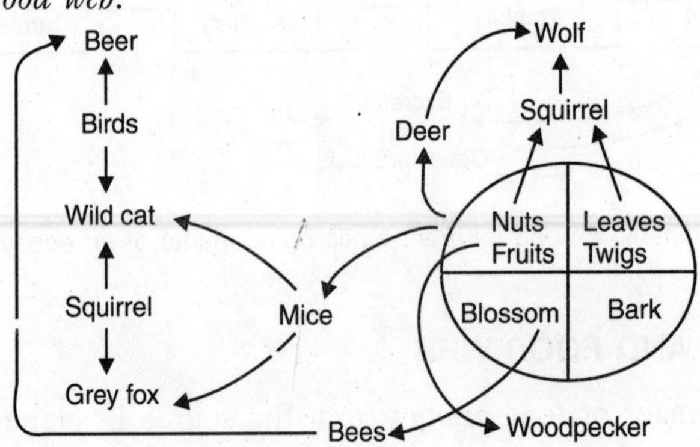

Fig. 3.6: Food web in terrestrial ecosystem

3.7 NATURAL/BIOGEOCHEMICAL CYCLES

Within an ecosystem there are dynamic relations between the living forms and their physical environment, i.e., rocks, air and soil of the earth (geo-). These relations are found as natural or biogeochemical cycles which involve continuous circulation of the essential elements and compounds required for life from environment to organisms and back to environment. The natural cycles and ecosystems function in a balanced manner which stabilises biosphere and sustains the life processes on earth.

The Geologic Cycle

The Geologic cycle means recycling of the earth's crust. According to the Plate Tectonic Theory, the earth's surface and interior processes are closely related. The lithosphere is subjected to continuous creation and destruction due to tectonic plate interactions giving rise to mountains in some places and depressions in some other places. It was initially composed of igneous or primary rocks due to cooling of magma and molten lava erupted from the interior of the earth. Furthermore, the earth's crust is constantly exposed to weathering forces viz., wind, rain and sun. These strip the land surface and carry sediments to the seas, oceans. The huge load of sediments carried to the seas and oceans gradually get cemented at the bottom and compacted to form sedimentary rocks.

The sequence: igneous rock–sedimentary rock–igneous rock (by flow of magma from the earth's core to the surface) is known as the Geologic Cycle as it is related to geologic processes.

The Hydrological Cycle

This cycle helps in exchange of water between air, land, sea, living plants and animals. About one-third of the solar energy absorbed by the earth is used to drive the hydrological cycle—massive evaporation of water from the oceans, cloud formation and rainfall which gives us our supply and reserves of fresh water. (see also p. 4).

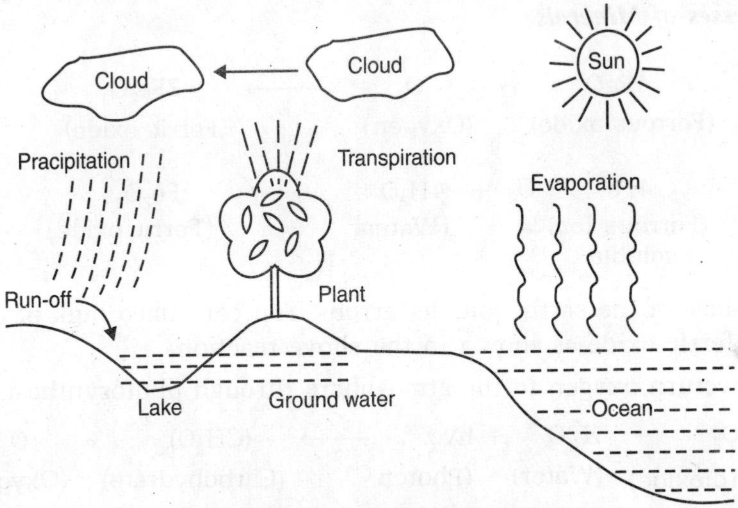

Fig. 3.7: The hydrological cycle

At freezing temperature rainwater freezes into snow and forms hail in presence of strong wind. Water as rain, snow and hail is precipitated on land and water surfaces. On land surfaces water seeps into the soil and is stored as ground water. The natural water level or water table exists below the ground. The water table is supported by the underlying clay and rock strata. Ground water does not remain static but moves in various directions. It moves up and reaches soil surface where it is drawn by plant roots.

Another important ground water resource is the aquifiers. This exists above impermeable rock strata – water percolates through porous rocks and forms these underground lakes or reservoirs. From the latter water can be pumped by digging tube wells and extracted by sinking wells.

The Oxygen Cycle

Oxygen is vital for life in the biosphere. It is a major component of all living organisms. It is drawn by man and animals during aerobic respiration from air but released by plants during photosynthesis setting up the oxygen cycle. The oxygen cycle is based on exchange of oxygen among the environmental segments—atmosphere, hydrosphere, lithosphere and biosphere. It plays an important role in atmospheric chemistry, geochemical transformations and life processes.

Combustion Reactions

$$\underset{\text{(Fossil fuel)}}{C} + \underset{\text{(Oxygen)}}{O_2} \longrightarrow \underset{\text{(Carbon dioxide)}}{CO_2}$$

$$\underset{\text{(Methane / natural gas)}}{CH_4} + 2O_2 \longrightarrow CO_2 + \underset{\text{(Water)}}{2H_2O}$$

Weathering Processes of Minerals

$$\underset{\text{(Ferrous oxide)}}{4FeO} + \underset{\text{(Oxygen)}}{O_2} \longrightarrow \underset{\text{(Ferric oxide)}}{2Fe_2O_3}$$

$$\underset{\text{(Ferrous ion) soluble}}{4Fe^{2+}} + O_2 + \underset{\text{(Water)}}{4H_2O} \longrightarrow \underset{\text{(Ferric oxide)}}{2Fe_2O_3}$$

In the early stage of the earth, soluble ferrous iron consumed bulk of oxygen giving large deposits of ferric oxide as shown in the above reaction.

Green plants return oxygen to the atmosphere through photosynthesis:

$$\underset{\text{(Carbon dioxide)}}{CO_2} + \underset{\text{(Water)}}{H_2O} + \underset{\text{(Photon sunlight)}}{h\nu} \longrightarrow \underset{\text{(Carbohydrate)}}{(CH_2O)_x} + \underset{\text{(Oxygen)}}{O_2}$$

The Nitrogen Cycle

Nitrogen and its compounds are essential for life processes in the biosphere. There is continuous exchange of nitrogen within the ecosystems operating the nitrogen cycle. Proteins produced by plants and animals in their metabolic processes are organic compounds of nitrogen. The major load of nitrogenous organic residue in soil originates from death and decay of plants and excreta of animals. These organic residues in soil are taken up by various soil micro-organisms for their metabolism which give products such as ammonia, nitrates and nitrites. Plants absorb nitrates from soil which re-enter the nitrogen cycle. Some soil micro-organisms break down soil nitrate into nitrogen by denitrification process while others transform nitrogen into soluble nitrogen compounds (see Fig. 3.8 The Nitrogen Cycle).

The Carbon Cycle

As carbon is the backbone of biological chemistry, the carbon cycle is a very important chemical cycle. The atmosphere is the minor reservoir of carbon dioxide while the oceans are the major reservoir, containing as much as 50 times more as that of air where it is stored as bicarbonate mineral deposit on the ocean floor. The latter regulates the carbon dioxide level in the atmosphere. The cycle operates in the form of carbon dioxide exchanging among the atmosphere, biosphere and the oceans (Fig. 3.8). The Carbon dioxide balance sheet per year is given:

(i) emissions by fossil fuel 20 billion tonnes,

(ii) emissions by deforestation and changes in land use 5.5 billion tonnes,

(iii) uptake in the oceans 5.5 billion tonnes,

(iv) uptake by carbon dioxide fertilization, i.e., photosynthesis, 7.3 billion tonnes:

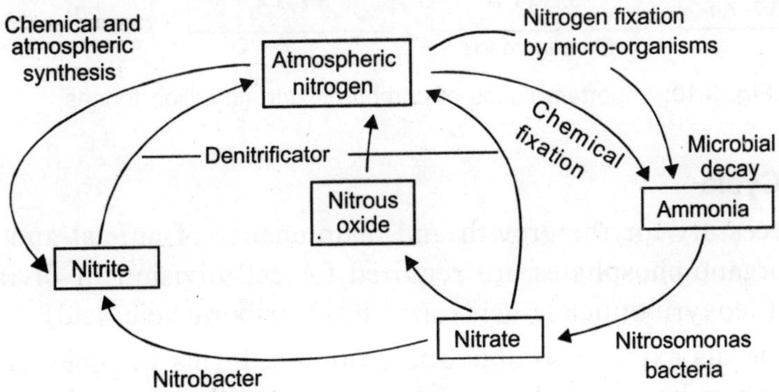

Fig. 3.8: The nitrogen cycle

Thus there is a net increase of carbon dioxide in the atmosphere of 11 billion tons per year. This can be reduced by 50 per cent if we can stop deforestation (Fig. 3.9).

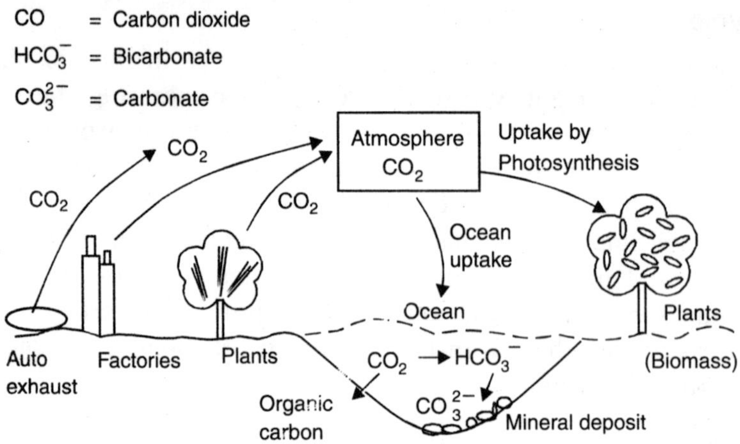

Fig. 3.9: Sources and sinks of carbon dioxide

The atmosphere contains 2700 billion tonnes of carbon dioxide; biosphere, vegetation and soil about 6600 billion tonnes and the oceans about 1,36,000 billion tonnes of carbon dioxide.

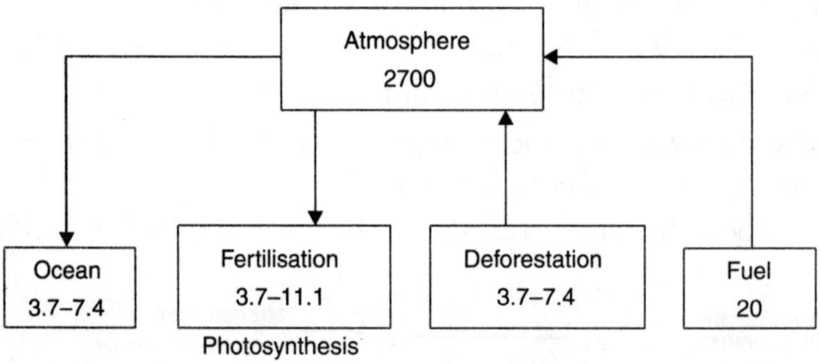

Fig. 3.10: Important fluxes of carbon dioxide (in billion tonnes)

The Phosphate Cycle

Phosphates are necessary for the growth and maintenance of animal and human bones and teeth while organo-phosphates are required for cell division involving production of nuclear DNA (deoxyribonucleic acid) and RNA (ribonucleic acid).

Phosphate minerals exist in soluble and insoluble forms in rocks and soil. Plants absorb inorganic phosphate salts from soil and change them into organic phosphate. Animals obtain their phosphate by eating plants. After death and decay, plants and animals return phosphates to the soil. Bulk of the phosphate in soil is fixed or absorbed on soil particles but part of it is leached out into water bodies.

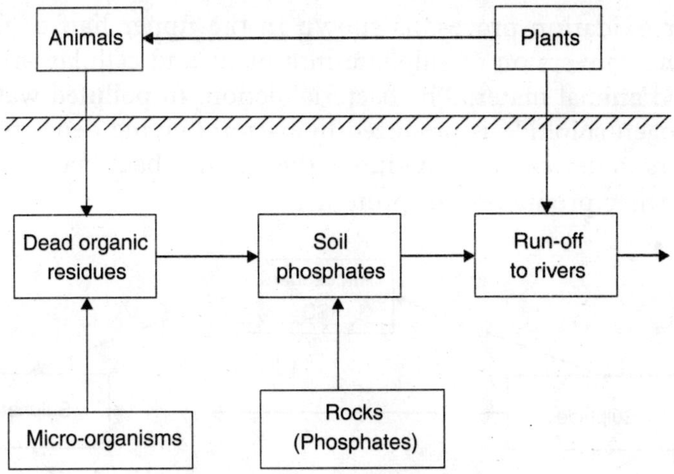

Fig. 3.11: The phosphate cycle on land

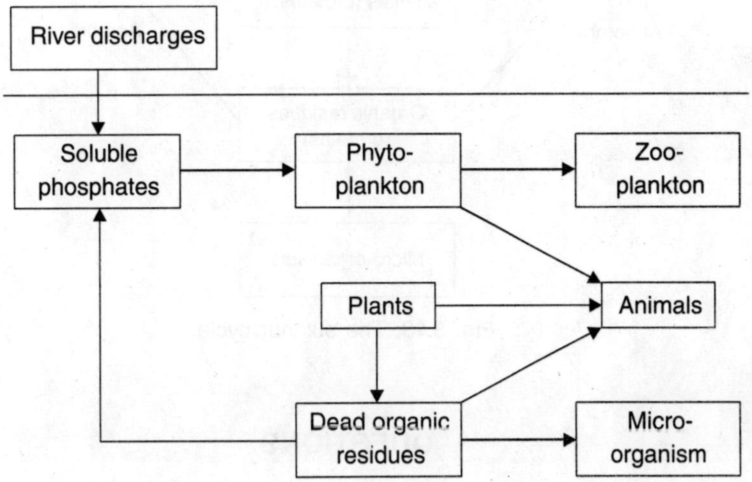

Fig. 3.12: The phosphate cycle in water

The natural phosphate cycle is affected by pollution, mainly from agricultural run-off containing superphosphate and also from domestic sewage. Phosphate pollution of rivers and lakes is the cause of algal bloom (eutrophication) which reduces dissolved oxygen in water and disrupts the food chain. The phosphate cycles on land and in water are shown in Figs. 2.10 and 2.11.

The Sulphur Cycle

Sulphur and its compounds are required by plants and animals for synthesis of some amino acids and proteins. Some sulphur bacteria act as the media for exchanges of sulphur within the ecosystems. The sulphur cycle (Fig. 2.12) illustrates the circulation of sulphur and its compounds in the environment.

The sulphur oxidation process is shown in the upper half of the cycle. The lower section shows the conversion of sulphate into plant and cellular proteins and the decay of dead plant and animal material by bacterial action. In polluted waters under anaerobic conditions hydrogen sulphide is produced by bacteria giving deposits of iron sulphide. In unpolluted waters under aerobic conditions the sulphur bacteria transform sulphides into sulphates for further production of proteins.

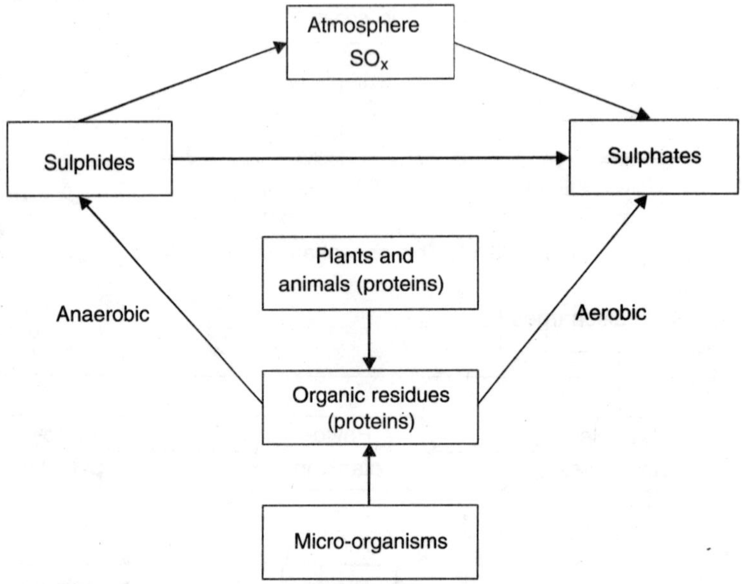

Fig. 3.13: The sulphur cycle

QUESTIONS

1. Define ecology and ecosystem.
2. Illustrate land ecosystem and also aquatic ecosystem.
3. What are the component of an ecosystem? Narrate their functions.
4. Discuss how the flow of energy in an ecosystem is related to food chain.
5. Illustrate food web in an ecosystem.
6. Comment on the role of Biogeochemical Cycles in Environment.
7. Write notes on:
   (a) Hydrological cycle (b) Carbon cycle
   (c) Nitrogen cycle (d) Oxygen cycle
   (e) Sulphur cycle

❍ ❍ ❍

4 Air Pollution and Control

4.1 ATMOSPHERIC COMPOSITION

[See Section 1.2/p.2–3: Atmosphere]

(1) Atmosphere: The atmosphere has three major constituents—major, minor and trace. Pure (i.e., pollution-free) dry air at ground level has the following components, expressed in *percentage by volume* (within brackets):

Major components:	Nitrogen (78.09)	Oxygen (20.94)	Water vapour (0.1)
Minor components:	Argon (0.9)	Carbon dioxide (0.032)	
Trace components:	Neon (0.0018)	Helium (0.0005)	Methane (0.0002) etc.

The properties of the atmosphere vary much with altitude. The density shows sharp decrease with increasing altitude. Pressure drops from 1 atmosphere at sea level to 3×10^{-7} atmosphere at 100 km above sea level while temperature varies from −92° to 1200° C. The total mass of the atmosphere is about 5×10^{7} tons which is roughly one millionth of the earth's total mass(5×10^{24} kg).

4.2 EARTH'S ENERGY / RADIATION BALANCE

The physical and chemical chat characteristics of the atmosphere and the critical heat balance of the Earth are determined by energy and mass transfer processes in the atmosphere.

The incident solar energy is mainly in the visible region of the spectrum. The shorter wavelength blue light is scattered relatively more strongly by molecules and particles in the upper atmosphere. This explains why *the sky is blue as it is viewed by scattered light*, Similarly, *light that has been transmitted through scattering in the atmosphere appears red, particularly around sunset and sunrise* and under circumstances in which the atmosphere contains a high level of particles.

The solar energy flux reaching the atmosphere is huge, amounting to 1.34×10^3 watts per square-metre (19.5 kcal per minute per square metre). This value is known as the solar constant and may be termed *insulation* which stands for '*incoming solar radiation*'. It all this energy reached the Earth's surface and were retained, the planet would have evaporated long ago. The complex factors involved in maintaining the earth heat balance within very narrow limits are crucial in retaining conditions of climate that will support life on earth. The great changes of climate that resulted in Ice ages during some periods or tropical conditions during others, were caused by variations of only a few degrees in average temperature.

The solar spectrum incident on the earth's atmosphere has a peak at 483 nm (visible region) indicating the surface temperature of the Sun at approx. 6000 K. The radiation emitted from the Earth's surface (earths spectrum) shows peak of about 10,000 nm (infrared region) with the surface temperature of the earth about 290 K.(17°C). In other words, the earth absorbs radiation mainly in the visible region but emits radiation in the infrared region (2–40μ with peak at 10μ).

The earth absorbs about 66% of the solar flux incident on it (19.5 kcal per sq. metre per minute) while it reflects or scatters back into space about 34% **(albedo)** of the solar flux.

Energy transport plays a key role in the Earth's radiation balance. This proceeds through the mechanism of (a) radiation of energy in the infrared region from the earth; (b) conduction of energy through the interaction of atoms or molecules, and (c) convection of energy through massive air circulation. The latter two mechanisms are responsible for loss of heat from the earth's surface through transport to clouds, and consequent radiation from the clouds.

The average surface temperature of the earth is maintained around 15°C due to re-absorption of most of the outgoing infrared radiation (2–40μ) by water vapour (4–8μ), and to a lesser extent by carbon dioxide (12–16.3μ). The combined effect of water vapour and carbon dioxide leads to the *greenhouse effect* which is of great significance in governing the climate on earth. This will be discussed in the following section.

Increasing agricultural and industrial outputs can also upset the earth's radiation balance by changing the *albedo* (fraction of sunlight reflected and scattered back to the atmosphere). Deforestation and the consequent soil erosion increase the *albedo*. Furthermore, both agriculture and industry release large quantities of dust and fumes into the atmosphere. The effects of these human practices are small compared to the natural forces (wind, sea, spray, volcanoes, etc.) which contribute to a large extent to the particle loading of the atmosphere. These atmospheric particles have a cooling effect on the earth's surface through increased scattering of the sun's rays. Dark particles (deposited on soot) can absorb light whereas light particles reflect light. The former helps in the heating of the earth's atmosphere, while the latter removes heat. The effect of particles on climate can be understood by a detailed analysis of the sources of these particles and their overall loading in the atmosphere (Sec. 2.9).

Albedo is important in determining earth's heat balance in that absorbed radiation heats the surface while reflected radiation does not. Albedo varies with the surface. For example, freshly precipitated snow has an albedo of 90% since it reflects 90% of incoming radiation while freshly plowed black topsoil (agricultural field) has an albedo of only 2.5%.

Temperature Inversion

During day time the ground is heated by solar radiation and air near the ground is heated by conduction from the ground. The packet of warm air rises until it reaches a region in the troposphere where the surrounding air has the same density i.e., same temperature and pressure. Air pressure falls due to upward movement leading to expansion of gas, which means doing work against intermolecular attraction for which energy is taken from air itself. In this process the energy and temperature of air drop. The expansion can be assumed to be adiabatic as no heat flows into or out of the packet during the process.

The rate of decrease of temperature with increase in altitude (height) is called the lapse rate. The average lapse rate is positive in the Troposphere. For dry air it is 1°C/ 100 metre while for water saturated air it varies with temperature and height due to release of latent heat as the air parcel rises and cools.

The region of increasing temperature with altitude is called *temperature inversion*. This exists at an altitude 400–600 m approximately. It is within this region that the heated air is trapped and smog as well as photochemical smog take place.

Temperature inversions are of two types—Radiation Inversion and Subsidence Inversion.

London tends to have Radiation Inversions in winter: the well known London smog of 1952 killed 2000 people. Los Angeles tends to have Subsidence Inversion that traps pollutants in summer and fall.

Radiation inversion generally occurs at night when air near the ground is cooled by conduction from the cooled surface, particularly on a clear night when radiation cooling occurs. Higher up in the troposphere, air is not cooled appreciably so that in absence of winds, there is *colder air underneath warmer air*, a case of inversion.

4.3 ATMOSPHERIC DISPERSION

Atmospheric dispersion is governed by several parameters viz., atmospheric temperature profile, smoke stack height, wind data, ambient temperature, adiabatic lapse rate, air quality standards, solar insulation, local terrain features, etc.

Temperature profile (Height-Temperature curve) and adiabatic lapse rate lead to maximum mixing depth and ventilation coefficient (Fig. 4.1). The maximum mixing depth is obtained by projecting dry adiabatic lapse rate to the point of intersection with atmospheric temperature profile. The product of maximum mixing depth and average wind speed within the mixing depth is termed *ventilation coefficient* (m^2 $sec.^{-1}$) Values of ventilation coefficient less than 6000 $m^2 sec^{-1}$ indicate high air pollution potential.

It is important to understand the dispersion of air pollutants from various emission sources. This will be illustrated with a specific example, smoke stack. Figure 4.2 shows different types of plume behavior from a smoke stack.

(*a*) *Coning*—Pluming emitted from a smoke stack into a neutral stable atmosphere forms symmetrical pluming.

(*b*) *Looping*—This kind of pluming occurs when atmosphere is very unstable and there is vertical air movement up and down.

(*c*) *Fanning*—This occurs in stable atmosphere which restricts dispersion of plume in the vertical direction.

(*d*) *Fumigation*—This results from a smoke stack that is under inversion layer with head downward much more easily than upward. It leads to largely elevated down wind, ground level concentrations.

(*e*) *Lofting*—The stack is above an inversion layer where mixing in the upward direction is unrestricted but downward motion is restricted by the inversion's stable air, causing relief to the people at the ground level.

A common strategy to air pollution control is to build taller and taller stacks so as to limit pollutants above inversion layer. In general, the height of stack should be 2.5 time of the height of surrounding buildings. However, there is the risk of smoke travelling great distances (as in case of acid rain) causing environment damage at faraway places.

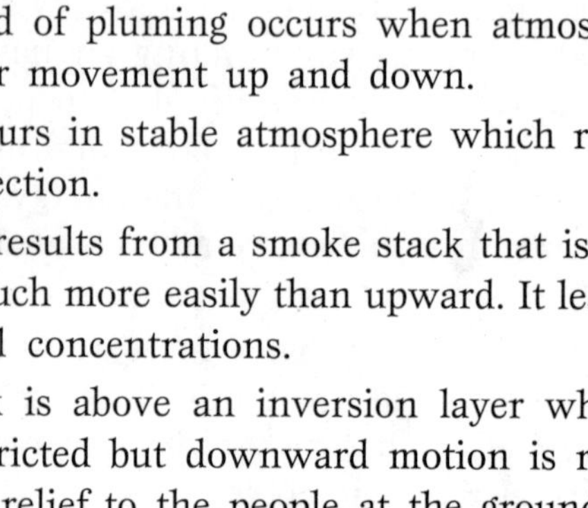

Fig. 4.1: Maximum mixing depth is obtained by projecting the dry adiabatic lapse rate to the point of intersection with atmospheric temperature profile.

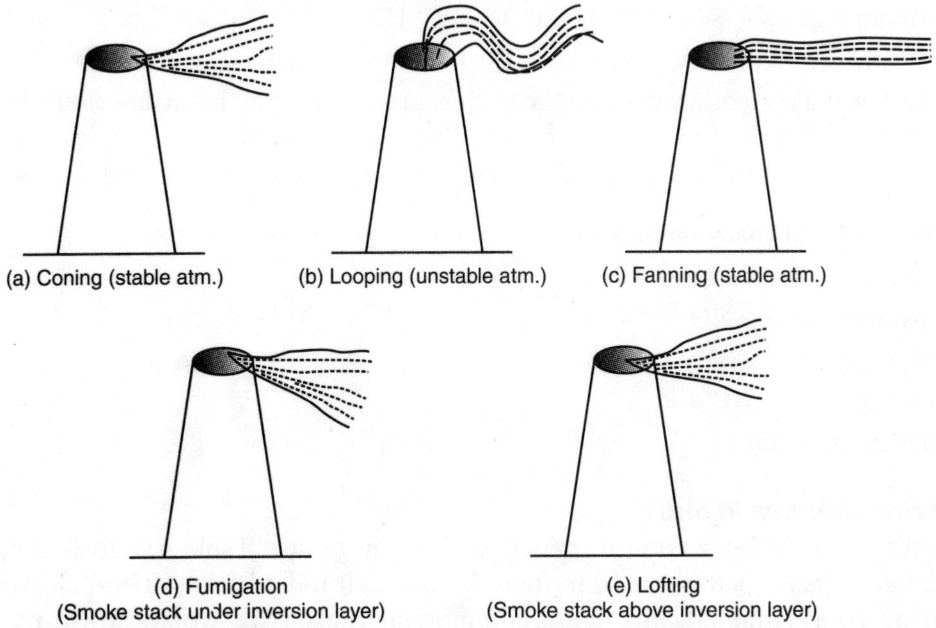

Fig. 4.2: Smoke stack plume behaviour under different atmospheric conditions

4.4 AIR POLLUTION

Pure air is colourless and odourless. But various pollutants from natural and man-made sources are entering the atmosphere daily and these disturb the dynamic equilibrium in the atmosphere. This leads to air pollution when the normal properties of air are upset and both man and environment suffer.

Natural sources of air pollution are:

- Volcanic activity, vegetation decay, forest fires emitting carbon monoxide, sulphur dioxide and hydrogen sulphide and tiny particles of solids or liquids sprayed from the seas and land by wind.

Man-made sources are:

- Gases, mists, particulates and aerosols emitted by industries and other chemical and biological processes used by man.

Air Pollutants

In general, air pollutants, emitted from natural and anthropogenic sources, can be broadly classified under two categories —Primary and Secondary pollutants. The primary pollutants are those that are emitted directly from the sources. These are:

Inorganic gases – SO_2, NO, CO, CO_2, H_2S, HF.

Olefinic and Aromatic hydrocarbons; Radioactive materials.

The Secondary pollutants are those that are formed in the atmosphere by chemical reactions among primary pollutants and atmospheric constituents. Examples are: SO_3, NO_2, PAN (Peroxyacetyl nitrate), O_3, aldehydes, ketones, various nitrate and sulphate salts.

The five major primary pollutants are:

Carbon monoxide CO

Nitrogen oxides, NO_x

Sulphur dioxide, SO_2

Hydrocarbons, HC and

Particulate matter.

Emission Sources in India

Air pollution originates from our fuel combustion policy. Table 4.2 gives the emissions of the five primary pollutants from fuels alone on all India basis (1976–77). Air pollution resulting from combustion of non-commercial fuels (fire wood, cowdung, etc.), far surpasses that from combustion of commercial fuels (coal, petroleum, etc.) However, the latter is the major source of air pollution in urban areas.

Table 4.2: Nationwide Air Pollutant Emission from Fuel Sources in India* (1976–77)

	Particulate matter	*SO_2*	*NO_x*	*CO*	*Organics including HC*
	(millions of tonnes)				
Commercial (coal, petroleum, natural gas)	3.21	2.15	0.62	1.29	2.80
Non-commercial (firewood, dry cattle dung, vegetable wastes, refuse burning)	7.54	4.67	1.97	2.05	6.19

In addition to burning of fuels, major industries like steel, paper and pulp, textiles, cement, and sulphuric and nitric acid plants contribute relatively small but significant amounts of air pollutants to the nation's atmosphere. Since most of the industries are located in major cities, these significantly add to the pollution burden of the metropolitan areas. Table 4.3 lists the various sources and their contributions to the pollution of a typical metropolitan city like Mumbai.

* Recent data (2006–07) are much higher (>50%) than those in Table 4.2 and Table 4.3.

Table 4.3: Major Pollutant Emissions in Greater Mumbai Area (1976–77)

Source	*Particulate matter*	*SO_2*	*NO_x*	*CO*	*HC*
		(Tonnes per day)			
Domestic	19.0	2.7	3.3	7.0	2.0
Commercial & institutional	2.2	1.7	1.0	—	—
Industrial	14.0	236.0	35.7	111.0	11.6
Power generation	28.0	62.0	30.0	2.0	—
Transportation	30.0	40.6	27.7	263.0	37.4
Total	93.2	343.0	97.7	383.0	81.0

As seen from the table, industry is the major source of pollution on a tonnage basis, followed by transportation and generation of electrical power. These three sources are responsible for the emission of about 77 per cent of particulate matter, 98 per cent of SO_2 and 94 per cent of NO_x.

Air pollution status of Indian metropolitan cities may be compared in this context.

According to UN Report in 1994 Kolkata, Mumbai and Delhi were among the most polluted cities in the world. During winter temperature inversion occurs when pollutants are trapped near the ground level.

Kolkata

Total vehicular emissions 300 tonnes/day; suspended particulate matter (SPM) 2000–3000 $\mu g/m^3$ (WHO limit 150–250 $\mu g/m^3$); Polynuclear aromatic hydrocarbons (due to automotive emissions and smoke from chullahs) 150–500 $\mu g/m^3$; SO_x 100–250 $\mu g/m^3$; NO_x 150–250 $\mu g/m^3$.

Delhi

SPM 150–700 $\mu g/m^3$

SO_x 10–35 $\mu g/m^3$

NO_x 70–180 $\mu g/m^3$

This shows that Kolkata air is more polluted than Delhi air.

In USA air pollution pattern (as on 1976 and valid even in 1995) shows that transportation accounts for more than 46% of total pollutants produced per year and hence remains the principal source of air pollution. Carbon monoxide is the major individual pollutant with a tonnage matching that of all the pollutants taken together.

Table 4.4: Primary Air Pollutant Sources and their Quantities (Million Tonnes per Year)

Sources	***CO***	***Weight of pollutants produced*** NO_x	***HC***	SO_x	***Particulates*** *<20μ>*	***30μ***	***Total weight of pollutant produced***
Transportation	70	10	10.8	0.8	1.2	1.0	94
Fuel combustion (stationary sources)	1.2	11.8	1.4	21.9	4.6	1.3	42.2
Industrial processes	7.8	0.7	9.4	4.1	6.3	2.7	31.0
Solid waste disposal	7.8	0.8	1.6	0.1	1.1	—	11.2
Miscellaneous	8.5	0.4	6.3	0.1	1.3	—	16.6
Total weight of pollutant produced (in million tonnes)	95.0	23.6	29.5	27.0	—	19.5	194.6

1μ =10^{-6} metre (1 part in 1 million parts of 1 metre)

* See Appendix 4: Toxic Chemicals in Air

Primary Pollutants

Carbon Monoxide, CO

It is a colourless, odourless and tasteless gas which is injurious to our health. Each year 350 million tonnes of CO (275 million tonnes from human sources and 75 million tonnes from natural sources) are emitted all over the world in which USA alone shares 100 million tonnes. Transportation accounts for 70 per cent of CO emission. That is to say, diesel and petroleum engines in automobiles are primarily responsible for about 70 per cent of CO emissions.

The sources of carbon monoxide, CO are the chemical reactions:

(i) incomplete combustion of fuel or carbon containing compounds:

$$\underset{\text{(Carbon)}}{2C} + \underset{\text{(Oxygen)}}{O_2} \longrightarrow \underset{\text{(Carbon monoxide)}}{2CO}$$

(ii) reaction of carbon dioxide and carbon-containing materials at elevated temperatures in industries e.g., in blast furnaces:

$$\underset{\text{(Carbon dioxide)}}{CO_2} + \underset{\text{(Carbon)}}{C} \longrightarrow 2CO$$

(iii) dissociation of carbon dioxide at high temperatures:

$$\underset{\text{(Carbon dioxide)}}{CO_2} \rightleftharpoons \underset{\text{(Carbon monoxide)}}{CO} + \underset{\text{(Oxygen)}}{O}$$

Sinks

Part of carbon monoxide is lost in the upper atmosphere. The major sink is soil micro-organisms. A potting soil sample weighing 28 kg completely removed in 3 hours 120 ppm carbon monoxide from ambient air. The same soil sample on sterilisation failed to remove carbon monoxide from air.

Control of CO Pollution

The petroleum and diesel-fed automobiles account for major share of carbon monoxide emission. Hence efforts for carbon monoxide pollution control are mainly aimed at automobiles. Use of catalytic converters in the internal combustion engines of automobiles helps in cleaning up the exhaust emissions. Such converters built into the automobile engines promote oxidation-reduction cycles and ensure complete combustion of carbon monoxide, nitrogen oxides and hydrocarbons. The following figure and flow-sheet illustrate the action of catalytic converters. Use of catalytic converters in two stages helps in elimination of pollutants from exhaust gases before they are discharged into the atmosphere.

In the first converter nitrogen oxides are reduced to nitrogen (+ ammonia) in presence of finely divided catalyst platinum, and the reducing gases, carbon monoxide and hydrocarbons. The production of ammonia is kept at a minimum under carefully controlled conditions. In the second converter, air is introduced to provide an oxidizing atmosphere for complete oxidation of carbon monoxide and hydrocarbon into carbon dioxide and water in presence of finely divided platinum catalyst.

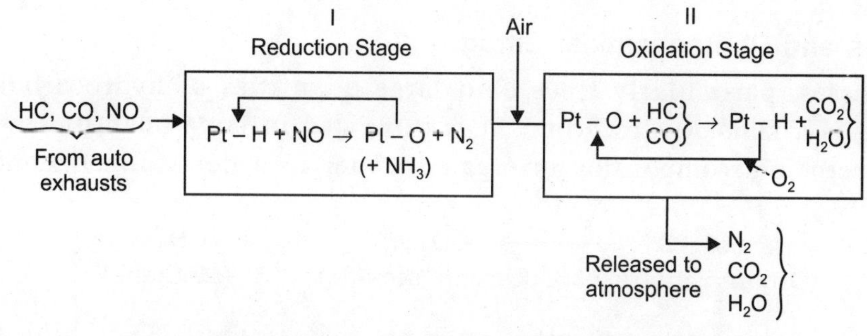

Fig. 4.3: Catalytic converters for treating auto emissions

Thus by means of platinum catalytic converters, auto exhaust emissions are cleaned up through reduction-oxidation reactions. In all developed countries it is mandatory by law for all automobiles to fit their engines with catalytic converters. In India some automobile companies have plans to fix their automobile engines with catalytic converters.

Nitrogen Oxides, NO_x

It consists of mixed oxides, nitric oxide and nitrogen dioxide (NO and NO_2 respectively)—the former is a colourless and odourless gas but the latter (NO_2) has a reddish brown colour and pungent smell.

The formation of NO and NO_2 is based on the chemical reactions:

$$\underset{\text{(Nitrogen)}}{N_2} + \underset{\text{(Oxygen)}}{O_2} \xrightleftharpoons{1210°-1765°C} \underset{\text{(Nitric oxide)}}{2NO}$$

$$2NO + O_2 \rightleftharpoons 2NO_2$$

These reactions occur inside the automobile engines so that the exhaust gases consist of NO_x. The latter concentration in rural air is much less than in urban air.

In air NO_x is converted into nitric acid, HNO_3 by natural processes:

$$\underset{\text{(Nitrogen dioxide)}}{NO_2} + \underset{\text{(Ozone)}}{O_3} \longrightarrow \underset{\text{(Nitrogen trioxide)}}{NO_3} + O_2$$

$$NO_3 + NO_2 \longrightarrow \underset{\text{(Nitrogen pentoxide)}}{N_2O_5}$$

$$N_2O_5 + \underset{\text{(Water vapour)}}{H_2O} \longrightarrow \underset{\text{(Nitric acid)}}{2HNO_3}$$

This nitric acid is one of the constituents of acid rain discussed in a subsequent section. From auto exhaust emissions NO_x is removed as discussed above by means of catalytic converters.

Hydrocarbons and Photochemical Smog

Natural processes, particularly trees emit large quantities of hydrocarbons in air. Methane, CH_4 is a major hydrocarbon. It is generated in large quantities by bacteria formed by anaerobic decomposition of organic matter in water sediments and soil.

$$\underset{\text{(Organic matter)}}{2\{CH_2O\}} \xrightarrow{\text{bacteria}} \underset{\text{(Carbon dioxide)}}{CO_2 \uparrow} + \underset{\text{(Methane)}}{CH_4 \uparrow}$$

Domestic animals (cattle, buffaloes, etc.) contribute about 85 million tonnes of methane to the atmosphere each year. Automobiles are significant sources of hydrocarbons.

In presence of ozone, carbon monoxide, nitrogen oxides and hydrocarbon participate in photochemical reactions (in presence of sunlight). A chain reaction proceeds in which the free radical R $\dot{C}H_2$ is generated in the first step. Other free radicals which are formed are: R $CH_2\dot{O}_2$ in the second step by reaction with oxygen, R CH_2 O; R $CH_2\dot{O}$ in the third step by reaction with nitric oxide; $H\dot{O}_2$ in the fourth step by reaction with oxygen; a stable aldehyde R CHO is another product at this stage; $H\dot{O}$ is formed in fifth step by reaction with nitric oxide (nitrogen dioxide is another product here); and finally, the starting free radical R $\dot{C}H_2$ is regenerated by reaction with hydrocarbon, R CH_3 thereby sustaining the chain reaction.

The harmful products in the chain reaction are NO_2 and aldehyde, R CHO. A side reaction also follows by another route through the aldehyde, R CHO; it gives an injurious end product, peroxyacetyl nitrate (PAN) which is a strong eye irritant. These reactions lead to photochemical smog formation, which is characterised by brown hazy fumes which irritate the eyes and lungs and also cause serious damage to plants.

Photochemical smog occurs in coastal cities in winter climate e.g., in Los Angeles, USA which have the heaviest vehicular traffic.

Sulphur Dioxide, SO_2

Sulphur dioxide is a colourless gas with a pungent odour. It is produced from the combustion of any sulphur-bearing material.

Sulphur dioxide, SO_2 is always associated with a little of sulphur trioxide, SO_3.

$$\underset{\text{(Sulphur)}}{S} + \underset{\text{(Oxygen)}}{O_2} \longrightarrow \underset{\text{(Sulphur dioxide)}}{SO_2}$$

$$2SO_2 + O_2 \longrightarrow \underset{\text{(Sulphur trioxide)}}{2SO_3}$$

Man-made sources—coal-fired power stations and other industries contribute about 33 per cent of SO_x pollution while natural sources, viz., volcanoes provide about 67 per cent of SO_x pollution.

Soot particles containing metal oxides catalyse the oxidation of sulphur dioxide to trioxide.

$$\underset{\text{(Sulphur dioxide)}}{SO_2} + \underset{\text{(Ozone)}}{O_3} \longrightarrow \underset{\text{(Sulphur trioxide)}}{SO_3} + O_2 \quad \xrightarrow{H_2O} \quad \underset{\text{(Sulphuric acid)}}{H_2SO_4} \longrightarrow \underset{\text{(Aerosol droplet)}}{(H_2SO_4)_n}$$

$$SO_2 + \tfrac{1}{2}O_2 \xrightarrow{\text{soot}} SO_3 \longrightarrow H_2SO_4 \longrightarrow \underset{\text{Aerosol droplet}}{(H_2SO_4)_n}$$

The first reaction above occurs in presence of ozone and water vapour. The product, sulphuric acid is formed on aerosol (fine particle suspended in air as in smoke, fog, mist, etc.) droplet. Sulphuric acid is one of the constituents of acid rain.

In winter climate sulphur oxides from thermal power plants along with other gases leads to *smog* formation e.g., *London smog.* This is known as reducing smog in contrast with photochemical smog which is known as oxidising smog (consisting of hydrocarbons, nitrogen oxides and ozone). *London smog* (1952) is well known for its disastrous effect. Heavy smog (SO_2) conditions prevailed in London for five days which killed about 4,000 people. The causes of death were bronchitis, pneumonia, and other respiratory troubles particularly among aged people.

Control of SO_x Pollution

SO_x (sulphur oxides) from flue gases of industrial plants can be removed by means of chemical scrubbers. The flue stack gases are led through a bed of (slurry) of limestone, $CaCO_3$ (calcium carbonate) which absorbs sulphur dioxide quite efficiently.

$$\underset{\text{(Calcium carbonate)}}{2CaCO_3} + \underset{\text{(Sulphur dioxide)}}{2SO_2} + \underset{\text{(Oxygen)}}{O_2} \longrightarrow \underset{\text{(Calcium sulphate)}}{2CaSO_4} + \underset{\text{(Carbon dioxide)}}{CO_2}$$

The method is economical but the disposal of solid waste, calcium sulphate is a problem.

Alternatively, sulphur dioxide in aqueous solution is treated with citric acid salt and the resulting solution is exposed to a stream of hydrogen sulphide gas whereby sulphur is deposited. This sulphur can then be recovered and utilised.

Thermal power plants, major sources of man-made SO_x pollution, are normally constructed with tall chimneys to disperse the emissions over a wide area. This reduces the local problem but creates problems for far away areas through acid rains (see below).

Acid Rain

It has been described above that much of nitrogen oxides, NO_x and sulphur oxides, SO_x entering the atmosphere are transformed into nitric acid (HNO_3) and sulphuric acid (H_2SO_4) respectively. These combine with hydrogen chloride, HCl from HCl emissions (both by man-made and natural sources) and generate acidic precipitation, known as *acid rain.*

Acid rain is a major environmental issue as it badly damages the environment. It damages buildings and structural materials of marble, limestones, slate and mortar. These materials become structurally weak as calcium carbonate reacts with sulphuric acid to form soluble sulphate, which is leached out by rain water:

$$\underset{\text{(Limestone)}}{CaCO_3} + \underset{\text{(Sulphuric acid)}}{H_2SO_4} \longrightarrow \underset{\text{(Calcium sulphate)}}{CaSO_4} + CO_2 + H_2O$$

In Greece and Italy invaluable stones and statutes have been partly dissolved by acid rain. Besides these, acid rain damaged forests in Germany and lakes in Sweden and Canada. Acid rain originated from U.K. but far away in Sweden, it damaged some 8,000 lakes of which 4,000 are dead. Similarly, acid rain from USA damaged lakes and forests in Canada. In India, Taj Mahal is threatened by acid rain from Mathura refinery and other industries.

Control of Acid Rain

Acid rain can be checked if its constituents, sulphur dioxide and nitrogen oxide are controlled as discussed above.

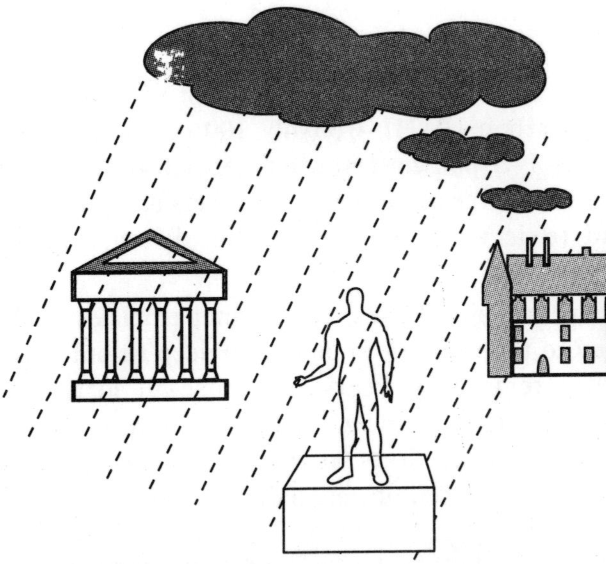

Fig. 4.4: Acid rain in Greece and Italy

Particulate

Small solid particles and liquid droplets are collectively termed *particulates.* They originate both from natural and man-made sources. Every year natural sources discharge 800–2,000 million tonnes and man-made sources 200–500 million tonnes of particulates. Among man-made sources, fly ash from thermal power plants deserve mention. The Table 4.5 gives a list of annual production of particulate matter from the two sources.

Table 4.5: World-wide Addition of Particulate Matter to the Atmosphere (in Million Tonnes)

	Annual Production (million tons)	
Particulate matter	***Natural source***	***Man-made sources***
Total particles	800–2000	200–450
Dust and smoke	–	10–90
Salt, forest fires	450–1100	
Sulphate	130–200	130–200
Nitrate	30–35	140–700
Hydrocarbons	15–20	75–200

Particulates range in size from 0.0002 μ (about the size of a molecule) to 500 μ (1 μ = 10^{-6} metre). The number of particles in the atmosphere vary from several hundred per cm^3 in clean air to more than 100,000 per cm^3 in highly polluted air (urban/ industrial area).

Soot

Soot particles originate from fuel combustion and consist of highly condensed product of polycyclic aromatic hydrocarbon (PAH)—roughly 100 condensed aromatic rings. The hydrogen content of soot is 1–3 per cent and oxygen content 5–10 per cent due to partial surface oxidation. Due to large surface area, soot acts as a carrier for toxic organics e.g., benzo-α-pyrene and toxic trace metals e.g., beryllium, cadmium, chromium, manganese, nickel, vanadium, etc.

A soot particle has an average size 0.1–20 μ. The finer particles (< 3 μ) are the worst causes of lungs damage due to their ability to penetrate deep in our respiratory tract and thence in lungs where they remain for years and cause all sorts of diseases such as cough, bronchitis, asthma, and finally cancer. Particulates cause increased corrosion of metals which assume serious dimensions in industrial and urban areas. They are responsible for damage to buildings, sculptures, paintings, etc.

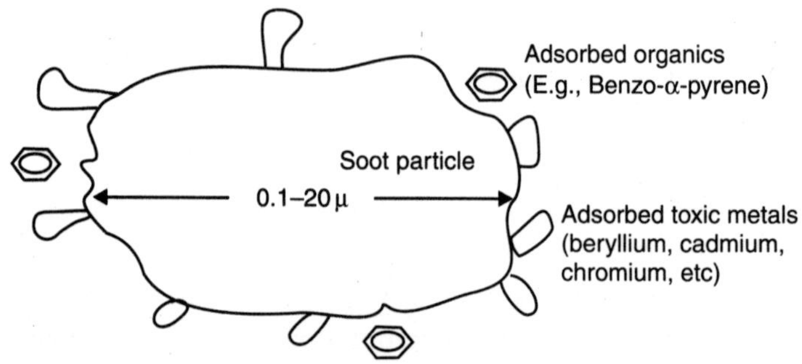

Fig. 4.5: A soot particle

Particulates play key roles in the atmosphere. They reduce visibility by scattering and absorption of solar radiation. They influence the climate through the formation of cloud, rain and snow by acting as nuclei upon which water can condense into raindrops. Atmospheric particulate levels can be correlated with the extent of precipitation over cities and suburbs.

Control of Particulate Emissions

The removal of particulate matter from gas streams is an essential step for air pollution control. There are four types of equipment used for this purpose.

(1) *Gravity settling chamber:* Effluent gases are led into a chamber which is large enough to permit gas velocities to decrease and dust or droplets to settle. The chamber is generally in the shape of a horizontal rectangular tank articles with

a diameter > 50 μ are ordinarily removed in this manner. The method is, however, not suitable for fine particles which require longer setting time.

(2) *Cyclone collector:* A gas flowing in a tight circular spiral produces a centrifugal force on suspended particles, forcing them to move outward through the gas stream to a wall where they are collected. Thus, it is possible to remove 95% particles in the diameter range 5–20 μ (Fig. 4.6).

(3) *Wet scrubbers:* These utilise a liquid (usually H_2O) to help remove solid liquid or gaseous contaminants. The extent of contact and interaction are increased by the use of spray chambers or towers where the liquid is introduced into the gas stream as fine spray.

(4) *Electrostatic precipitators:* These are based on the principle that aerosal particles acquire charges when subjected to an electrical field:

$$F = Eq$$

Where F = force in dynes, to which the particles are subjected,

E = voltage gradient (volt/cm)

q = electrostatic charge on the particles (esu).

The particles acquire a charge when a gas stream is led through a high voltage dc-corona. The charged particles are attracted to a grounded surface from which they are removed. Ozone is a possible product of the corona discharge.

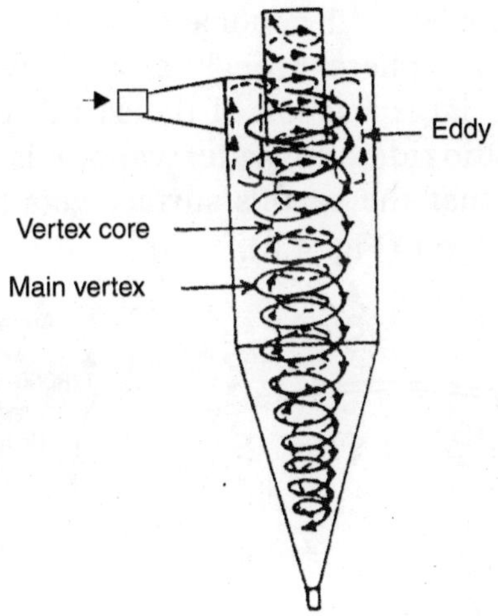

Fig. 4.6: Typical cyclone collector
[L.Byers, "Controlling Atmospheric Particulates," *Technology Tutor,* 1.46 (1971)].

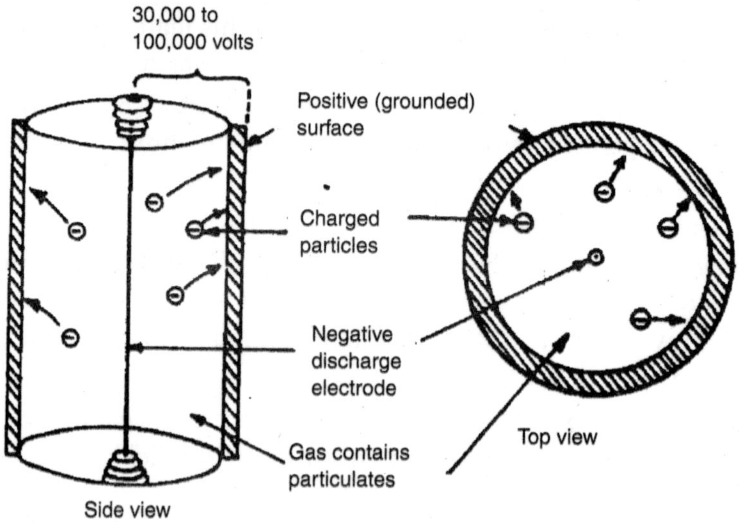

Fig. 4.7: Schematic diagram of an electrostatic precipitator

4.5 GREENHOUSE EFFECT (GLOBAL WARMING)

Carbon dioxide is a non-pollutant gas in the atmosphere and a minor constituent (356 parts per million) but it is of serious concern for the environment for its ability to change the global climate.

The earth's surface partly absorbs sun's rays while emits long-wave infra-red radiation (8000–25000 nanometres; 1 nanometre = 10^{-9} metre = 1 nm). Carbon dioxide and water vapour in the atmosphere strongly absorb infra-red radiation (14,000–25,000 nm) and effectively block a large fraction of the earth's emitted radiation. The radiation thus absorbed by carbon dioxide and water vapour is partly returned to the earth's surface. The net result is that the earth's surface gets heated and the phenomenon is known as the *Greenhouse Effect* (Fig. 4.8).

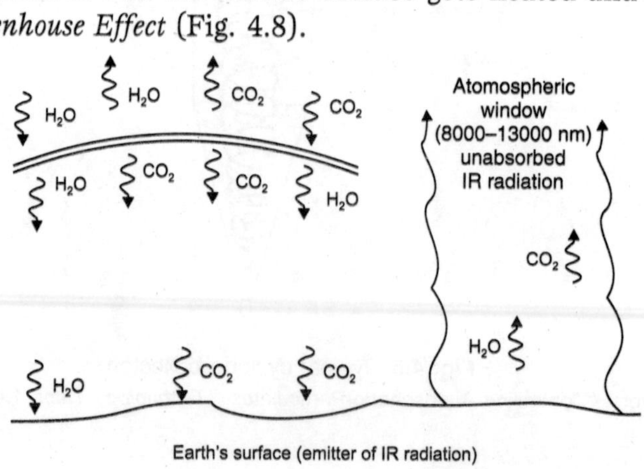

Fig. 4.8: The greenhouse effect

The carbon dioxide level in air has increased from 280 ppm (pre-Industrial revolution era 1760) to 350 ppm at present in two centuries. Fossil fuel (petrol, diesel, coal) combustion is the major source of increase of carbon dioxide level increasing at the rate of 1–2 per cent per year. At this rate of increase, the earth's surface temperature may rise as much as 2°C in the next 100 years. However, nature has its check and balance system. The rate of increase of carbon dioxide is only 50 per cent of its expected magnitude due to the sinks, viz., oceans and photosynthesis by green plants (Fig. 4.9).

It may be noted that a slight rise in temperature even by 1°C, can have adverse effect on the world food production. (1) Thus wheat producing zones in the northern latitude will be shifted from CIS (former USSR) and Canada to the north pole and in India from UP, Punjab, Haryana to the Himalayas, i.e., from fertile to non-productive soil. In other words, wheat production will badly suffer. (2) The biological productivity of the ocean will fall due to warming of the surface layer. This reduces transport of nutrient from deeper layers of the ocean to the surface by vertical circulation. Moreover, there will be less photosynthesis by marine plants. In other words, the production of sea food (marine plants and fish) will decline. Sea food constitutes more than 30 per cent of our food supply. (3) Another disastrous effect is the rise in sea levels by as much as 15 cm in the next 100 years due to partial melting of polar ice caps. This sea level rise would threaten coastal cities (Kolkata, Mumbai, Chennai, etc.) and some 60-odd island nations such as Maldives, Bangladesh, etc. which will be drowned under the sea. When Himalayan snow melts and gets exhausted, the Himalayan rivers including Ganga may dry up —the Ganga valley will be hot and north India will lose its population base. At the same time India will lose its major life line, Ganga.

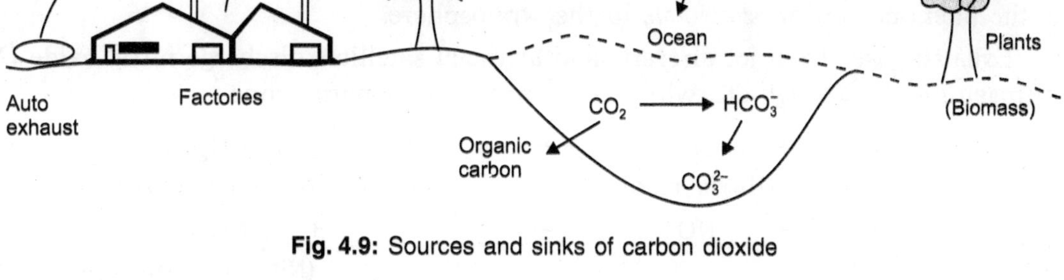

Fig. 4.9: Sources and sinks of carbon dioxide

Other Greenhouse Gases

Carbon dioxide is not the only culprit responsible for Greenhouse Effect and global warming. Other greenhouse gases are: methane, chlorofluorocarbons, nitrous oxide, ozone and water vapour. The relative contributions of these gases to greenhouse effect are:

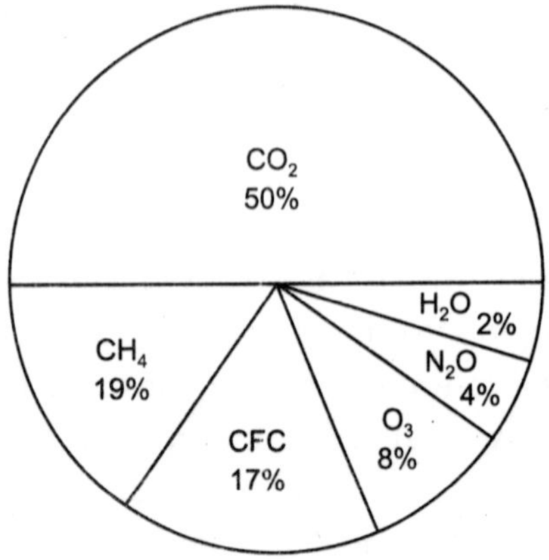

Fig. 4.10: Greenhouse gases (relative shares)

Carbon dioxide 50 per cent; Methane 19 per cent; Chlorofluorocarbons 17 per cent; ozone 8 per cent; nitrous oxide 4 per cent; water vapour 2 per cent.

This shows that carbon dioxide accounts for 50 per cent of the greenhouse gases. The shares of methane (19 per cent) and chlorofluorocarbons (17 per cent) (gases from refrigerators and air conditioners) cannot be ignored.

4.6 OZONE HOLE

In the stratosphere, the second region of the atmosphere, ozone is present in small quantities but it is protective shield for the earth. Ozone strongly absorbs ultraviolet radiation from the sun (295–320 nm) which is injurious for life on earth. Thus it protects living species on earth. But recent human activities have injected some dangerous chemicals in the stratosphere which consume ozone and reduce its concentration. This is the phenomenon of *ozone hole* in the stratosphere.

Exhaust gases from jet aircrafts and artificial satellites discharge nitric oxide (NO), nitrogen dioxide (NO_2), etc. which immediately react with ozone.

$$\underset{\text{(Ozone)}}{O_3} + \underset{\text{(Nitric oxide)}}{NO} \longrightarrow \underset{\text{(Oxygen)}}{O_2} + \underset{\text{(Nitrogen dioxide)}}{NO_2}$$

$$O_3 + NO_2 \longrightarrow O_2 + \underset{\text{(Nitrogen trioxide)}}{NO_3}$$

Chlorofluorocarbons (*CFC*) are used as coolants in refrigerators and air conditioners. These slowly pass from troposphere and stratosphere and once there, they stay for 100 years.

In presence of ultraviolet radiation (200 nm) from the sun, CFC breaks up into chlorine-free radical (Cl) which readily consumes ozone.

$$CFCl_3 \xrightarrow[\text{Ultraviolet radiation}]{hv} CFCl_2 + \dot{Cl}$$

$$\dot{Cl} + O_3 \longrightarrow Cl\dot{O} + O_2$$

$$Cl\dot{O} + O \longrightarrow \dot{Cl} + O_2$$

The free radical ($\dot{Cl}$) is regenerated and continues the chain reaction. It is estimated that one molecule of CFC consumes one lakh molecules of ozone. The damage by CFC continues for 100 years. Even if CFC production is stopped now all over the world, the CFC that is already there in the stratosphere will continue to damage the ozone layer for the next 100 years.

In 1979 ozone hole was observed in the sky over Antarctica—here ozone layer thickness was reduced by 30 per cent. Later on ozone hole was discovered in the sky over the thickly populated northern hemisphere. Here in winter ozone thickness was reduced by 4 per cent and in summer by 1 per cent.

Ozone hole allows passage of ultraviolet radiation to the earth where it causes skin cancer, eyesight defect, genetic disorder, etc. in the biosphere (man, animal and plant). In Europe and USA there is an increase in the cases of skin cancer among people while some million people are suffering from eye cataract.

In Montreal Conference (Montreal Protocol, 1987) and London Conference (1992) it was decided that the developed countries would totally ban CFC production by 2000 and the developing countries by 2010 A.D. But as stated above, even after the ban is enforced, the CFC and Cl shall continue their havoc for another 100 years. Researches are on for development of CFC-substitutes as coolants for refrigerators and air conditioners.

El Nino

Greenhouse effect, as discussed above, is an atmospheric warming phenomenon while El Nino is an ocean warming phenomenon. El Nino is a Spanish term for the Christ child as it starts in December in South American coast but it affects climate over half the globe.

Under normal conditions the water of the Eastern Pacific off Ecuador, Peru and Chile is cooler by 10°C than the waters of the Western Pacific. This promotes fisheries along the Eastern Pacific coast since cold waters, rich in nutrients, spring from the deep layers of the ocean. The trade winds blow along the equator from Southeast Pacific towards West pushing warm waters out to the ocean. But once every three to five years the trade winds reverse their direction, i.e., from west to east thereby pushing warm waters to the east. As a result, the Eastern Pacific water warms up by about 4°C (24° to 28°C) which disrupts the anchovy fishery, key to the Peruvian economy. This happens when El Nino appears. It not only kills fish but warms the air which lowers the atmospheric pressure and sparks storms and heavy rainfall along Chile coast and spreading up to Southern California (USA).

In 1982 El Nino raised sandstorms in Australia, cyclone in Tahiti, caused droughts in Africa and floods along the Californian coast in USA. Recently in 1998 El Nino activity has been felt all over the world. In the east coast of USA, winter was missing in January–February and in the west coast there was record rainfall in 100 years of US history. Research is on to understand the science of El Nino so that it can be forecast and advance warning signal sent to the regions where lives and properties can be saved in time.

4.7 AIR POLLUTION AND BIOSPHERE

Air pollutants can be transported to the earth's surface by atmospheric turbulence where they interact with the ocean surface, vegetation and upper layers of the soil and are removed by absorption or chemical reaction. Absorption at these surfaces depends on many factors about which little information is available for many pollutants.

The primary pollutants discharged into the atmosphere suffer chemical changes in presence of H_2O, O_2 and ultraviolet ray of sunlight to form secondary pollutants. Typical examples are:

$$CO_2 + H_2O \longrightarrow H_2CO_3 \text{ Carbonic acid}$$
$$SO_2 + H_2O + O_2 \longrightarrow H_2SO_4 \text{ Sulphuric acid}$$
$$NO_2 + H_2O \longrightarrow HNO_3 \text{ Nitric acid}$$
$$H_2S + O_2 \longrightarrow H_2SO_4 \text{ Sulphuric acid}$$

These pollutants have overall effects on soil, vegetation, crops, animals, man and materials.

Plants are affected by gaseous pollutants and deposition of particulates on soil. Acid rain over a period of time tends to lower soil pH and makes it acidic. Furthermore, deposition of toxic metals on soil in industrial areas render the soil unsuitable for growth of plants. Some plants are very sensitive to traces of toxic metals as the latter inhibit the action of plant enzymes. On the other hand, particulates such as dust and soot, deposited on plant leaves, block the stomata of plants. This in turn effects the plants by restricting the absorption of CO_2 and reducing the rate of photosynthesis and also the rate of transpiration. The net result is retarded growth of plants and decreased yield of crops. In Leeds (U.K.) there was drastic reduction in growth of lettuce and raddish in heavily polluted (SO_2) areas compared to the less polluted area of the city.

O_3(Ozone) and PAN (Peroxyacetyl nitrate) attack plants by oxidising their–SH groups of proteins into disulphides. This oxidation results in inhibition of individual enzyme activity.

Cattle are also affected by air pollution. In 1957 Smithfield cattle show (U.K.) was held under heavy smog conditions. After the show was over, cattle developed breathing troubles and many died while sheep and pigs were not so affected.

Corrosion of metals – SO_2 polluted air accelerates corrosion of metals. Iron and its alloys and Zn, Cu, Al used in outdoor structures as well as marble sculptures are vulnerable to corrosion. Salt spray in seaside air also corrodes iron structures in buildings in the coastal areas.

Meteorology and Air Pollution

Air pollution, one of the man-made activities, has some impact on meteorology, i.e., the science of atmospheric phenomenon. Meteorology is based on the physical parameters such as temperature, wind, moisture and movement of air masses in the atmosphere. It is also affected by the chemical properties of the atmosphere and the chemical reactions going on in the atmosphere.

The air pollutants get dispersed in the atmosphere depending on the patterns of air circulation. In this context *temperature inversion* plays an important role. It occurs when a warm air mass moves above a cold air mass resulting in air stagnation of the latter (cold air) in which air pollutants get trapped. The air above the ground becomes polluted. This happens when warm air blows over a mountain range and over cool air on the other side of the range. Such a phenomenon is observed in Denver, USA on the east of the Rocky mountains.

Human activities are partly responsible for changing the meteorology of the earth. These activities are:

1. Deforestation and loss of forest cover;
2. Shifting of surface water and ground water in large amounts;
3. Release of heat from power plants;
4. Emission of particles and trace gases into the atmosphere;
5. Release of carbon dioxide into the atmosphere by combustion of fossil fuels;
6. Emission from transport system into the lower and upper atmosphere.

The loading of particulate matter into the atmosphere influences the climate. It has been shown in Table 4.2 that each year natural sources are injecting about 800–2,000 million tonnes while man-made sources about 200–400 million tonnes of particles in the atmosphere. As already discussed, particles induce cloud formation and rainfall. It has been calculated as an approximation that if the particle loading increases by 50 per cent, the average temperature of the earth will decrease by about 0.5°C to 1°C due to particle-induced cloud formation. This partly counterbalances the temperature rise due to Greenhouse Effect.

Weather is determined by physical parameters of the atmosphere (temperature, humidity, wind speed and movement of air masses in the atmosphere). It is inter-related with climate. Climate system consists of the atmosphere, hydrosphere, lithosphere and biosphere i.e., all the components of the environment. It is a dynamic regime subject to changes on time scales from decades to millennia. The type of climate is governed by latitude and longitude of a country (Tropical winter etc.).

4.8 AIR QUALITY STANDARDS

Each pollutant, present in air, has a *threshold limit value* (TLV) which, if exceeded, causes public health hazards. Table 4.6 gives a list of typical pollutants with their threshold limits (TLV). For factory workers TLV sets the limit of exposure for 40-hour week (8 hours a day) without adverse effects. These TLV values are determined mainly by experiments on animals.

Table 4.6: Threshold Limit Values (TLV) for Some Common Pollutants (Gases and Vapours)

Pollutant	*Threshold Limit*	*Values*
	ppm	*mg/m³*
Acetone	750	1780
Ammonia	25	18
Arsenic	0.2–0.5	—
Benzene	10	20
Cadmium	—	0.05
Carbon dioxide	5000	9000
Carbon monoxide	50	50
Carbon tetrachloride	5	30
Chlorine	10	30
Chloroform	10	50
Hydrogen chloride	5	7
Hydrogen sulphide	10	14
Lead	—	0.2
Nitric oxide	25	30
Ozone	0.1	0.2
Sulphur dioxide	2.0	5.0
Vinyl chloride	5	10

4.9 MONITORING METHODS FOR AIR POLLUTANTS

The monitoring methods for some principal primary pollutants will be described.

Analysis of Sulphur Dioxide

The modified *West-Gaeke spectrophotometric method* remains the standard method for monitoring of 0.0005–5 ppm SO_2 in ambient air. The method was developed by West and Gaeke and later optimised. Figure 4.11 illustrates the sampling train employed to draw samples from air. The various components involved in such sampling are shown.

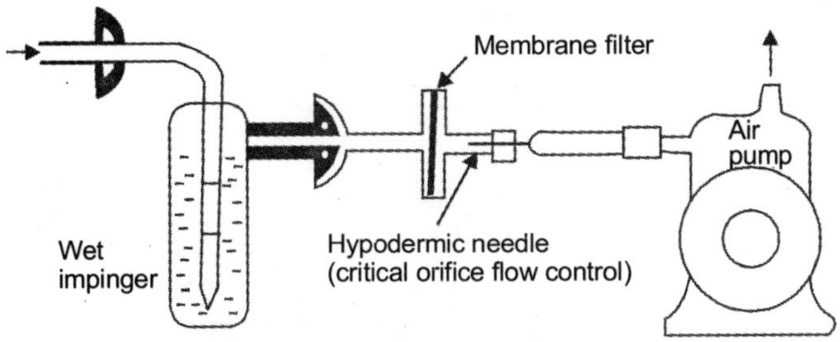

Fig. 4.11: Sampling train for the collection of sulphur dioxide from the atmosphere

SO_2 is collected in a scrubbing solution containing $HgCl_4^{2-}$ $(HgCl_2 + KCl)$, the collection efficiency being around 95%. The solution is allowed to react with HCHO and then with para-rosaniline hydrochloride:

$$HgCl_4^{2-} + SO_2 + H_2O \rightarrow HgCl_2SO_3^{2-} + 2H + 2Cl^-$$

$$SO_2 + H_2O + HCHO \rightarrow HOCH_2 — SO_2H$$

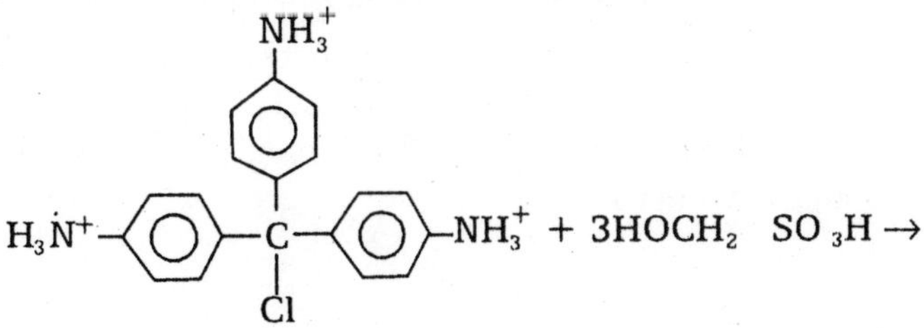

Colourless
(in acidic medium)

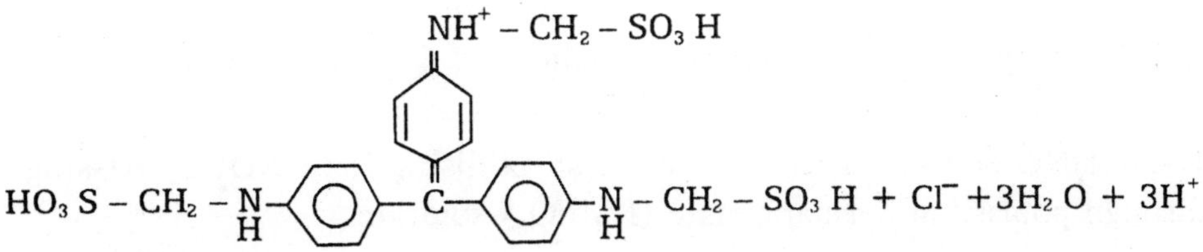

red voilet dye

The absorbance of the product red-violet dye is measured at 548 nm. A block diagram of a continuous SO_2 monitor, based upon the West-Gaeke method* and developed by Technicom in the form of the Technicom Air Monitor IV instrument, is shown in Fig. 4.12.

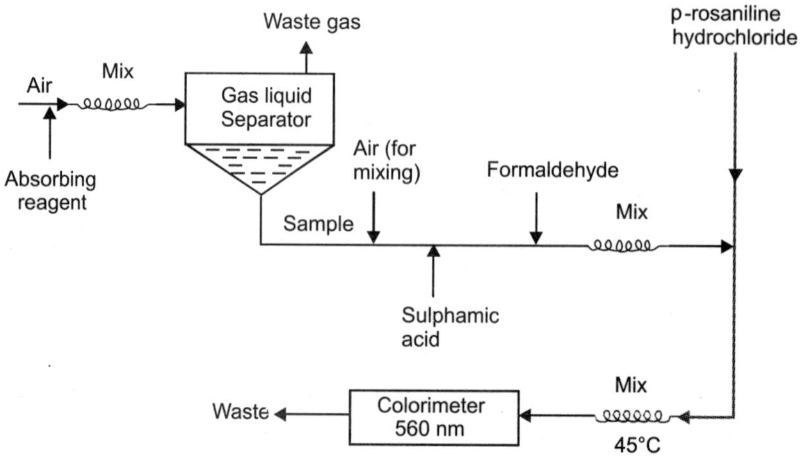

Fig. 4.12: Sulphur dioxide monitor using a modified West-Gaeke method

The major interference from NO_2 (> 2 ppm), when present, is eliminated by the addition of sulphamic acid, H_2NSO_3H. This reagent acts as a reducing agent, converting NO_2 to N_2.

Conductimetry is one of the common methods for automatic monitoring of SO_2 in industries. The sampled air containing SO_2 is passed through a dilute solution of H_2O_2 in dil. H_2SO_4

$$SO_2 + H_2O_2 \rightarrow H_2SO_4$$

The increase in conductivity is proportional to the concentration of SO_2 in the sample. With this technique as low as 0.01 ppm of SO_2 can be detected. However, the method is subject to interference from acidic gases such as hydrogen, chloride (+ve error) and ammonia (–ve error).

Nitrogen Oxides

Colorimetric method–This is an excellent method—known as Griess-Saltzmann method—for measurement of nitrogen oxides. It is based on reduction of NO_2 with sulphanilic acid to form diazonium salt. The latter is coupled with N-(1-naphthyl) ethylene diamine dihydrochloride to get a pink-coloured dye complex. Colour development is complete within 15 mins, at room temperature and measured at 550 nm.

Analysis of NO can be carried out by first oxidising it to NO_2 by passing the air sample through potassium permanganate ($KMnO_4$) solution.

Chemiluminescence Method

Chemiluminescene is the standard method of monitoring of NO_2. The phenomenon results from the emission of light from electronically excited species arising out of a chemical reaction. NO reacts with O_3 to produce electronically excited NO_2 molecule, which in the next step loses energy and returns to the ground state through emission of light (600–3000 nm). The latter is measured by a photomultiplier, the intensity being proportional to the concentration of NO_2 (Fig. 4.13):

$$NO + O_3 \rightarrow NO_2^{\bullet} + O_2$$

$$NO_2^{\bullet} \rightarrow NO_2 + h\nu$$

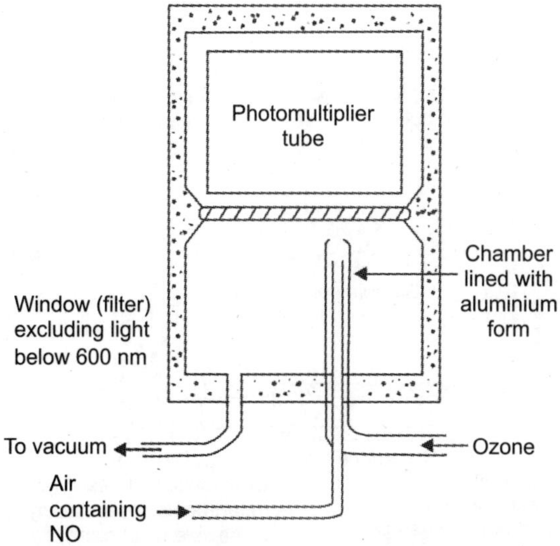

Fig. 4.13: Chemiluminescence detector for NO

NO_2 is invariably associated with NO in air sample. It is, therefore, necessary to convert NO_2 to NO before analysis. This can be effected by passing the air sample over a thermal converter which brings about the required chemical reaction. Analysis of another sample which is not led through thermal converter gives NO. The NO_2 concentration can be calculated from the difference between these two data.

Other nitrogen compounds, besides NO and NO_2, interfere if present in large excess. Particulates also interfere—however, their interference can be eliminated by using a membrane filter at the air inlet.

The method is also applicable to analysis of O_3 by interaction with NO.

Carbon Monoxide

Non-dispersive infrared spectrometry has been introduced for the analysis of CO. When IR radiation is led through a long cell (100 cm) containing CO, part of the energy of the radiation is absorbed by CO, which can be correlated with the concentration of CO.

In the non-dispersive IR spectrometer the IR radiation from the source is not dispersed according to wavelength by a prism or grating as in a standard IR spectrometer. It is made very specific for a given compound or type of compound by using the material under investigation as part of the detector. Radiation from an IR source is 'chopped' by a rotating device so that it alternately passes through a sample cell and a reference cell (Fig. 4.14). Both beams of light are incident on a detector which is filled with CO gas and divided into two compartments by a flexible diaphragm. The relative amounts of IR radiation absorbed by CO in the two sections of the detector are determined by the CO level in the sample. The difference in the amount of IR radiation absorbed in the two compartments give rise to temperature differences so that the diaphragm, bulges slightly to one side and this can be measured with a relative accuracy of 5% in the optimum concentration range.

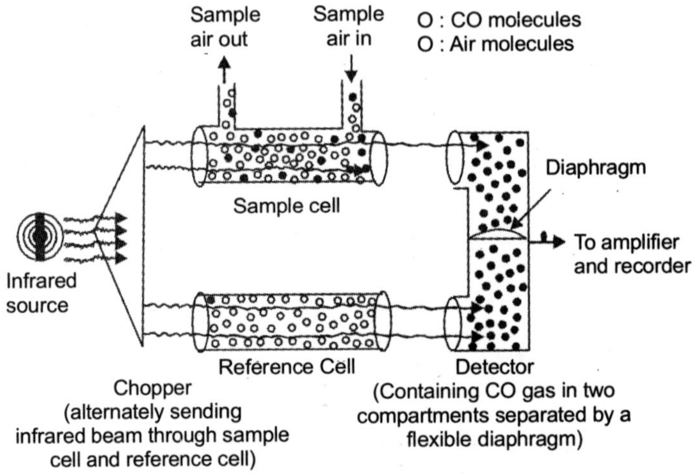

Fig. 4.14: Non-dispersive infrared analyzer for monitoring carbon monoxide

Hydrocarbons

Gas chromatography–Hydrocarbon pollutants are conveniently estimated in a gas chromatograph (GC) using flame ionization detector. As little as 0.1 μg of hydrocarbon/m^3 of air can be estimated by this technique.

Gas chromatography is based on the principle that when a mixture of volatile materials, transported by a carrier gas, is led through a column containing an absorbent solid phase or, more commonly, an absorbing liquid phase coated on a solid material, each volatile component will be partitioned between the carrier gas and the solid or liquid. Depending on the retention times in the column, the volatile component will emerge for the column at different times and finally be detected by a suitable detector. This permits both qualitative and quantitative analyses.

The block diagram of a gas chromatography is shown in Fig. 4.15 and that of a typical gas chromatography in Fig. 4.16.

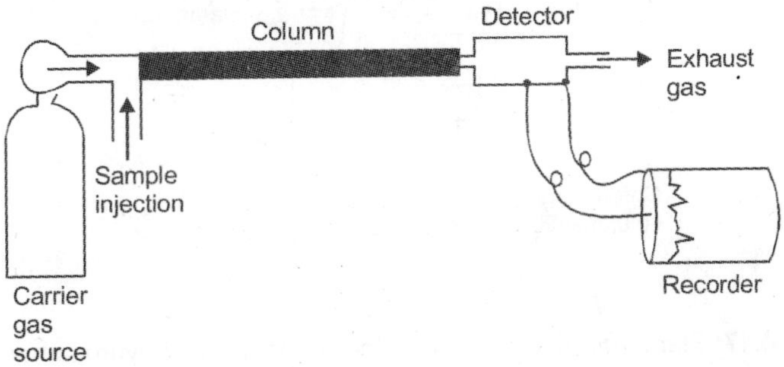

Fig. 4.15: Schematic diagram of gas chromatography

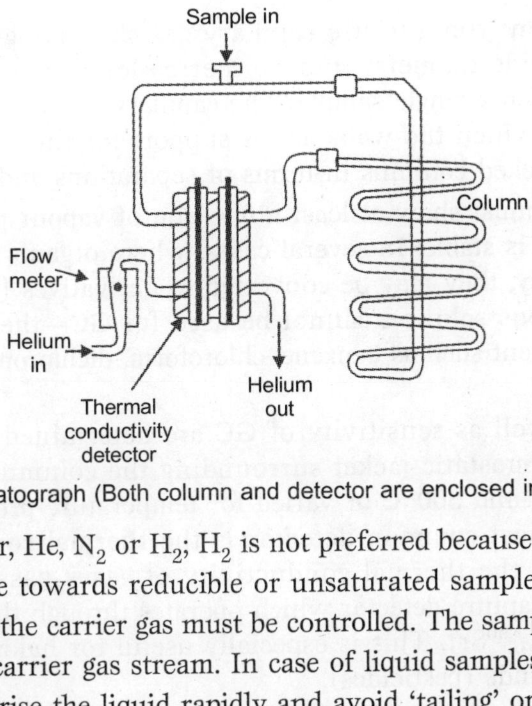

Fig. 4.16: A gas chromatograph (Both column and detector are enclosed in a thermostat chamber)

The carrier gas is Ar, He, N_2 or H_2; H_2 is not preferred because of its fire and explosion hazards and its reactive towards reducible or unsaturated sample components. The flow rate (60–80 ml/min) of the carrier gas must be controlled. The sample is injected as a single compact plug into the carrier gas stream. In case of liquid samples, the injection chamber must be heated to vaporise the liquid rapidly and avoid 'tailing' or spreading of the peaks.

In gas-solid chromatography (GSC), the fixed phase consists of a solid material such as granular SiO_2, Al_2O_3 or C. The more common type is gas-liquid chromatography (GLC) in which the fixed phase is a non-volatile liquid (silicon oil, polyethylene glycol, etc.) held as a thin layer on a solid support (diatomaceous earths or crushed firebrick, 60–80 mesh or 100–200 mesh or size). The packed column is usually 4 mm (i.d.) tubing of stainless steel or glass either bent in a V-shape or coiled. Lengths of such tubes are from 120 cm to 150 m. (Fig. 4.17).

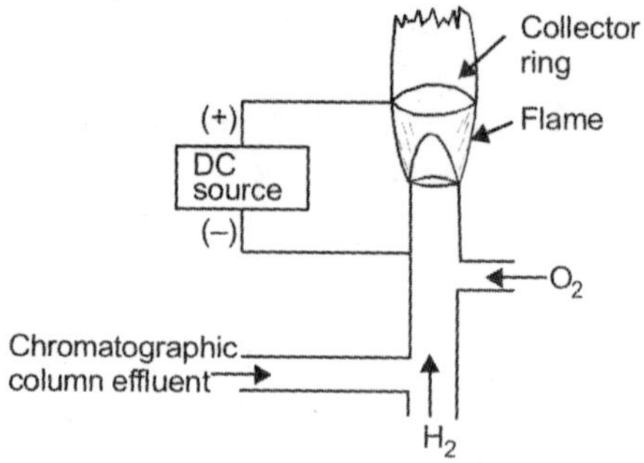

Fig. 4.17: Flame-ionization detector for the analysis of hydrocarbons.
(Reprinted by permission of Brook/Cole Publishing Company, Monterey, California 93940, Manahan, p. 235, 1979. Willard Grant Press, Statler Office Building, Boston, Massachusetts)

It has been the recent trend to use capillary gas chromatography where capillary columns (0.25 mm inside diameter and 50 metres long) can resolve up to several hundred components from a single sample. The capillary column may be of metal, glass or organic polymer, in which the walls act as support for the stationary liquid phase. They are superior to packed columns in terms of separations and efficiency.

For GC, the sample must show at least a few mm of vapour pressure at the highest temperature at which it is stable. In several cases, where organic compounds cannot be chromatographed directly, they may be converted to derivatives (e.g., esters) which are amenable to GC. Aqueous solution cannot be used for GC—the components must be brought into organic solvent such as benzene, chloroform, dichloromethene and hexane by solvent extraction.

The selectivity as well as sensitivity of GC are determined by the detector. The temperature of the thermostatic jacket surrounding the column and detector may be maintained between 50° and 300°C or varied for temperature programming, depending on requirement. The most common detector is the thermal conductivity cell which responds to changes in the thermal conductivity of gases passing over it. Another detector is the electron capture detector which operates through the capture of electrons emitted by a source (e.g., ^{90}Sr). This is especially useful for halogenated hydrocarbons and phosphorus compounds (pesticides).

The *flame ionization detectors* (FID) is extremely sensitive for the detection of organic compounds. In a flame, organic compounds form highly conducting fragments, such as C^+. On applying a potential gradient across the flame, a small current is generated which is readily measured. The response of a flame ionization detector is roughly proportional to the number of carbon atoms in the compound. Halogens, nitrogen or oxygen, if present in the organic compounds, decrease the sensitivity of the detector. The FID has wide range of linear response and is sensitive to very low levels of hydrocarbons, ranging down to 10^{-3} μg.

QUESTIONS

1. What are the major regions of the atmosphere? State their respective altitudes and temperature ranges. Name the important chemical species in these regions. Give your answer with a table and figure only.
2. Name the major, minor and trace constituents of the atmosphere. Give their percentage composition by volume.
3. "Carbon dioxide is a non-pollutant but it is of much environmental concern." Explain with reference to the Greenhouse Effect.
4. What is Greenhouse Effect? Describe its impact on global climate, food production and world geography.
5. What are the Greenhouse gases? What are their relative contribution to Greenhouse Effect?
6. What are the primary air pollutants? Describe their sources, and state their annual production figures (in million tonnes).
7. How would you control carbon monoxide and nitrogen oxides in automobile exhaust emissions?
8. How will you control sulphur dioxide and particulate emissions from thermal power plants?
9. Write explanatory notes on:
   (a) Ozone hole
   (b) El Nino
   (c) Acid rain
   (d) Photochemical smog
   (e) Effect of air pollution on public health
   (f) Effect of air pollution on meteorology
   (g) Air quality standards.
   (h) Atmospheric dispersion

❍❍❍

5 Water Pollution

> **"...The Indian philosophy of "*Vasudhaiva Kutumbakam*" which means the whole Universe is one family must dominate global efforts to protect the global commons",...**
>
> —R.K. PACHAURI, Nobel Peace Prize Address (2007).

5.1 WATER POLLUTION

The normal uses of water for public supply are—recreation (swimming, boating, etc.), fish, other aquatic life and wild life, agriculture (irrigation), industry, navigation, etc. Any change in the dynamic equilibrium in aquatic ecosystem (water body/biosphere/atmosphere) disturbs the normal function and properties of pure water and gives rise to the phenomenon of water pollution. The symptoms of water pollution of any water body/ground water are:

- Bad taste of drinking water,
- Offensive smells from lakes, rivers and ocean beaches,
- Unchecked growth of aquatic weeds in water bodies (eutrophication),
- Dead fish floating on water surface in river, lake, etc.
- Oil and grease floating on water surface.

The quality of water is of vital concern for mankind since it is directly linked with human welfare. It is a historical fact that faucal (human excreta or stool) pollution of drinking water caused water-borne diseases which wiped out entire populations of cities. In the developing countries like India, every day some 25,000 people die of water-borne diseases e.g., jaundice, hepatitis, cholera, dysentery, etc. In India about 2 lakhs out of 6 lakh villages have no access to safe drinking water—women have to walk 1–14 km daily for collecting water for drinking and cooking. In urban areas 40 per cent people are without access to safe water. The major sources of water pollution are domestic sewage from urban and rural areas, agricultural run-off (wash water) and industrial waste which are directly or indirectly discharged into water bodies.

Water Pollutants

The large number of water pollutants are broadly classified under the categories:

1. Organic pollutants,
2. Inorganic pollutants,
3. Sediments,
4. Radioactive materials and
5. Thermal pollutants.

Organic Pollutants

These include domestic sewage, pesticides, synthetic organic compounds, plant nutrients (from agricultural run-off), oil, wastes from food processing plants, paper mills and tanneries, etc. These reduce dissolved oxygen (D.O.) in water. Dissolved oxygen (D.O.) is essential for aquatic life, the optimum level being 4–6 ppm (parts per million). Decrease in D.O. value is an indicator of water pollution. The organic pollutants consume D.O. through the action of bacteria present in water.

Sewage and agricultural run-off provide plant nutrients in water giving rise to the biological process known as *eutrophication.* Large input of fertiliser and nutrients from these sources leads to enormous growth of aquatic weeds which gradually cover the entire waterbody (*algal bloom*). This disturbs the normal uses of water as the water body loses its D.O. and ends up, in a deep pool of water where fish cannot survive.

The production of synthetic organic chemicals (more than 60 million tonnes each year since 1980) multiplied more than 10 times since 1950. These include fuels, plastic fibres, solvents, detergents, paints, food additive, pharmaceuticals, etc. Their presence in water gives objectionable and offensive tastes, odour and colours to fish and aquatic plants.

Oil pollution of the seas has increased over the years, due to increased traffic of oil tankers in the seas causing oil spill and also due to oil losses during off-shore drilling. Oil pollution reduces light transmission through surface water and hence reduces photosynthesis by marine plants, decreases D.O. in water causing damage to marine life (plants, fish, etc.) and also contaminates sea food which enters the human food chain.

Pesticides have been largely used for killing pests and insects harmful for crops and thereby boosting the crop production. At present there are more than 10,000 different pesticides in use. They include insecticides (for killing insects) e.g., DDT (dichloro diphenyl trichloroethane), herbicides (for killing weeds and undesirable vegetation) and fungicides (for killing fungi and checking plant disease).

It has been found that pesticide residues contaminate crops and then enter the food chain of birds, mammals and human beings. The persistent pesticide, viz., DDT (which is not degraded in the environment) accumulates in food chain, getting magnified in each step from sea weed to fish and then to man by about ten thousand times (10^4). Thus the average level of DDT in human tissues is found to be 5–10 ppm, maximum being among Indians (25 ppm) compared to Americans (8 ppm).

Inorganic Pollutants

This group consists of inorganic salts, mineral acids, metals, trace elements, detergents, etc.

Acid mine drainage: Coal mines, particularly those which have been abandoned, discharge acid (sulphuric acid) and also ferric hydroxide into local streams through seepage. The acid on entering the waterbody destroys its aquatic life (plants, fish, etc.).

Sediments

Soil erosion, as a matter of natural process, generates sediments in water. Solid loadings in natural water are about 700 times as large as the solid loading from sewage discharge. Soil erosion is enhanced 5–10 times due to agricultural and 100 times due to construction activities. Bottom sediments in aquatic bodies (streams, lakes, estuaries, oceans) are important reservoirs of inorganic and organic matter, particularly trace metals e.g., chromium, copper, nickel, manganese and molybdenum.

Radioactive Materials

Radioactive pollution is caused by mining and processing of radioactive ores to produce radioactive substances, use of radioactive materials in nuclear power plants, use of radioactive isotopes in medical, industrial and research institutes and nuclear tests. The discharge of radioactive wastes into water and sewer systems is likely to create problems in future.

Thermal Pollutants

Coal-fired or nuclear fuel-fired thermal power plants are sources of *thermal pollution.* The hot water from these plants is dumped as waste into nearby lake or river where its temperature rises by about 10°C. This has harmful effect on aquatic life in the water body whose D.O. is reduced and as a result, fish kill is quite common.

When sewage and agricultural run-off containing phosphates or other nutrients enter natural waterbodies, they cause overnutrition, leading to *eutrophication*. The latter is the process of providing a waterbody with nutrients for the aquatic life it supports. A lake starts its lifecycle as *oligotrophic*, i.e., a clear body of water. With the introduction of nutrients through land run-off and growth and decay of aquatic life, the lake collects a good amount of organic substances. Evantually, there is algal bloom when the lake becomes marsh or debris. The stage is *eutrophic*, when the lake is filled with sediment while aquatic animal life will perish. It will then turn into dry land.

The rate of eutrophication strikes a balance between the production of aquatic life and its destruction by bacterial decomposition. With large inputs of nutrients from human sources, bacterial decomposition cannot keep pace with productivity and sedimentation is accelerated whereby eutrophication is favoured. Lakes can be protected from eutrophication only by providing measures for sewage treatment and preventing the sewage from entering the waterbodies.

Ground Water Pollution/Arsenic Contamination

Ground water is relatively free from surface contamination as it is located more than about 50 ft. below the land surface and the surface water gets filtered or screened by the underlying layers of soil, sand and stone pieces. But even then it gets contaminated due to leaching of minerals below the earth's surface.

An important case is that of *Arsenic (As)-contamination of ground water.* This arises from excessive pumping of ground water by shallow tube wells for irrigation in some West Bengal districts along the Hooghly river course and also in Bangladesh along the Padma river course. In this process air (oxygen) is injected into ground water bed which leaches the overlying mineral, iron pyrites (iron, arsenic, sulphide), oxidises it and releases arsenic into ground water.

More than one million people in six districts of West Bengal drink arsenic-contaminated ground water from tube wells in the region. Among them 20 lakh people suffer from various diseases related to arsenic poisoning like loss of hair, brittle nails, bronchitis, gangrene, etc. Several hundred deaths have also been reported. Similar calamity has threatened the lives of Bangladesh in the districts along the Padma river course.

Arsenic-biochemical Effects

Arsenic (III) exerts its toxic action by attacking—SH groups of an enzyme, thereby inhibiting enzyme action.

By virtue of its chemical similarity to P, as interferes with some biochemical processes involving P. As (III) compounds at high concentration coagulate proteins, complex with co-enzymes and affect phosphorylation. In general, As (III) is more toxic than As (V) and has carcinogenic effect.

Speciation of Arsenic

The species of As (III) in natural water bodies are As (III), As (V), methyl arsenic acid, dimethyl arsenic acid; in air, As (III), + As (V), dimethyl arsenic acid and trimethyl arsine.

The permissible dose of As in domestic water supplies is 0.05 ppm or 50 mg/litre. It is interesting to note that in delicious crabmeat dish total As content is 10 ppm or more. But in crab meat As is tied up largely as arsenobetaine, a strong complex, which is not toxic and hence harmless.

Drinking Water Supply

Treatment of drinking water supply is a matter of public health concern. The water treatment plants, in general, are simpler than sewage treatment plants. They operate in three steps—(i) aeration to settle suspended matter; (ii) coagulation of small particles and suspended matter by lime and ferric chloride; (iii) disinfection by chlorination to kill viruses, bacteria, etc. The purified water is then supplied by municipalities through pipes for domestic uses.

Industrial Waste Water Treatment

Industrial wastes contain toxic chemicals which can damage environment (water, soil, air) much more than domestic sewage. These waste liquids (effluents) can be purified by filtration using activated charcoal or ion exchange resins. Activated charcoal has large surface area and is an effective filter medium for adsorption of organic molcules. Synthetic organic ion exchange resins are very useful for removal of industrial waste metals (cations) and non-metals (anions).

Water Quality Standards

The analyses required of water samples depend on the intended use of the water. For example, if its intended use is drinking, water should meet certain quality criteria with respect to the appearance, (turbidity, colour), potability (taste, odour), health (bacteria, nitrates, chlorides, etc.) and toxicity (metals, organics). These and similar criteria are established by health or other regulating agencies to ensure that the water quality in a resource is suitable for the proposed use.

Based on the criteria, quality standards are set, which reflect the current state of knowledge of various water constituents. These standards are continuously revised as more and more is learnt about the effects of water constituents on proposed uses. Hence, these standards should not be used as absolute limits, but only as guidelines that can be used for preliminary judgements. Table 5.1 summarises several quality criteria and their standards for drinking water as suggested by the following agencies:

(1) Indian Council of Medical Research (ICMR)
(2) World Health Organisation (WHO)
(3) United States Public Health Service (USPHS)

Standards have been prepared for raw waters to be used as a source for various industrial applications. The specific purpose for which the water is used usually controls the requisite water quality. Water used in food and beverage industries will need to meet standards similar to those for drinking water, while water for cooling purposes can contain much higher concentrations of impurities. Table 5.2 illustrates the water quality characteristics of raw surface waters that have been used as source for various industrial operations. The values listed are the maximum concentrations of different constituents in the raw surface waters used in the specified industry, and not the maximum concentrations that could be tolerated.

Table 5.1: Standards for Drinking Water

A – Recommended maximum concentration[a] (mg/l except where shown otherwise)
B – Maximum permissible concentration[b] (mg/l except where shown otherwise)

	ICMR		WHO		USPHS	
	A	B	A	B	A	B
Physical:						
Turbidity (unit)	5	25	5	25	5	—
Colour (unit)	5	25	5	50	15	—
Chemical:						
*pH, units	7–8.5	(6.5 or 9.2)	7–8.5	(6.5 or 9.2)	6.0–8.5	—
Total solids	—	—	500	1500	500	—
Calcium	75	200	75	200	100	
Magnesium	50	150	50	150	30	—
Iron	0.3	1.0	0.3	1.0	0.3	—
Manganese	0.1	0.5	0.1	0.5	0.05	—
Copper	1.0	3.0	1.0	1.5	1.0	—
Sulphate	200	400	200	400	250	—
Phenols	0.001	0.002	0.001	0.002	0.001	—
*Fluorides	1.0	2.0	0.5	1.0–1.5	1.5	—
*Nitrates	20	50	—	50–100	10	—
Toxic:						
Arsenic	—	0.2	—	0.2	0.01	0.05
Barium	—	—	—	—	0.05	1.0
*Cadmium	—	—	—	—	0.01	0.01
*Chromium	—	0.05	—	0.05	0.05	0.05
Cyanide	—	0.01	—	0.01	0.01	0.2
Lead	—	0.1	—	0.1	—	0.05
Selenium	—	0.05	—	0.05	—	0.01
Silver	—	—	—	—	—	0.05
Bacteriological:	—	10 coliform per 100 ml	—	10 coliform per 100 ml		10 coliform per 100 ml

[a] Constituents should not be present in excess of listed concentrations where other more suitable supplies are or can be made available.
[b] Constituents in excess of the concentrations listed shall constitute grounds for rejection of the supply.
* See also Appendix-3.

Table 5.2: Maximum Concentrations of Constituents in Raw Water Supplies for Various Industries, mg/l

Constituent	*Boiler water (0–1500) psig*	*Cooling water (once through)*	*Textile industry*	*Pulp and paper*	*Chemical industry*	*Petroleum industry*
Silica (SiO_2)	150	50	—	50	—	85
Aluminium	3	3	—	—	—	—
Iron	80	14	0.3	2.6	5	15
Manganese	10	2.5	1.0	—	2	—
Calcium	—	500	—	—	25	220
Magnesium	—	—	—	—	100	85
Ammonia	—	—	—	—	—	40
Bicarbonate (HCO_3)	600	600	—	—	600	480
Sulphate	1400	680	—	—	850	900
Chloride	19000	600	—	200	500	1600
Nitrate	—	30	—	—	—	8
Phosphate	—	4	—	—	—	—
Dissolved solids	35000	1000	15	1080	2500	3500
Suspended solids	15000	1500	1000	—	10000	5000
Hardness	5000	850	120	475	1000	900
Alkalinity	500	500	—	—	500	500
pH (units)	—	5.0–8.9	6.0–8.0	4.6–9.4	5.5–8.0	6.0–9.0
Colour (units)	1200	—	—	360	500	25
COD	100	—	—	—	—	—
Temperature (°F)	120	100	—	95	—	—

5.2 MONITORING TECHNIQUES AND METHODOLOGY

pH

pH is measured by a pH meter using a glass electrode which generates a potential varying linearly with the pH of solution in which it is immersed. It is a Nernstian concentration cell with potential controlled by the activities of H^+ on either side of a very thin glass membrane. The latter is the bottom part of a bulb at the end of a glass tube containing a reference solution of fixed a_H^+.

$$E = \text{Constant} + \frac{RT}{nF}\ln\frac{a_H^+\,(\text{Sample})}{a_H^+\,(\text{Standard})}$$

$$= \text{Constant} + 0.058 + \text{pH at } 20°\text{C}$$

A calomel or Ag/AgCl/KCl reference electrode is usually located around the glass electrode stem (Fig. 5.1) for sample operation.

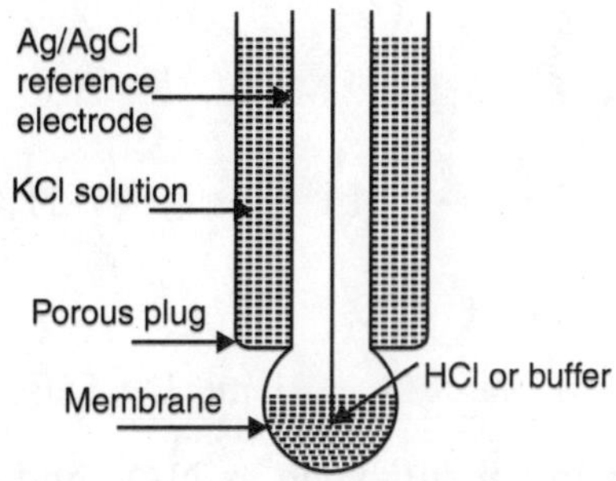

Fig. 5.1: A combined glass/Ag-Ag electrode

Procedure

1. Calibrate the electrode(s) with two standard buffer solutions of pH 4.0 and 9.2.
2. Rinse the electrode(s) thoroughly with deionised distilled water and carefully wipe with a tissue paper.
3. Dip the electrode(s) into the sample solution, swirl the solution and wait up to 1 minute for steady reading. A pH meter reading within ± 0.1 unit will be adequate for such work.

Specific Conductance

A simple conductivity meter with dip-type cell is used for this purpose. The instrument and cell are calibrated using 0.005M KCl solution (conductivity = 654 μmho cm^{-1}).

$$\text{Specific conductance, } k = \frac{1}{R} \cdot \frac{A}{I}$$

where R is the observed resistance of a column of electrolyte, 1 cm long and cross-section area, A cm^2.

Dissolved Oxygen (D.O.)

In natural and waste waters, D.O. levels depend on the physical, chemical and biological activities of the water body. The analysis of D.O. plays a key role in water-pollution control activities and waste-treatment-process control. Two methods are used, the Winkler or iodometric method and the electrometric method using a membrane electrode.

Winkler Method

D.O. is allowed to react with I^- to form I_2, which is then titrated with standard $Na_2S_2O_3$ solution. A fast quantitative reaction is ensured by addition of Mn(II) salts in strongly alkaline medium:

$$Mn^{2+} + \frac{1}{2}O_2 + 2OH^- \rightarrow MnO_2 \downarrow + H_2O$$

$$MnO_2 + 2I^- + 4H^+ \rightarrow Mn^{2+} + I_2 + 2H_2O$$

$$I_2 + 2S_2O_3^{2-} \rightarrow 2I^- + S_4O_6^{2-}$$

$$5 \text{ mL of } 0.025 \text{ M } Na_2S_2O_3 \equiv 1 \text{ mg L}^{-1} \text{ D.O.}$$

Interferences due to oxidizing agents such as NO_2^- and SO_3^{2-}, present in waste water, may be eliminated by addition of NaN_3 to alkaline I^- solution. On acidification NO_2 is decomposed.

$$N_3H + HNO_2 \rightarrow N_2 + N_2O + H_2O$$

Procedure

1. To a 100 ml sample in a 250 ml bottle, add 2 ml of 40% KF (to mask Fe^{3+}), 2 ml of 36% $MnSO_4$ and 2 ml of alkaline iodide-azide solution (50 g NaOH + 13.5 g NaI + 1.0 g NaN_3 diluted to 1 litre).
2. Shake well, allow the precipitate to settle, and then add 6 ml of 12 NH_2SO_4.
3. Shake the mixture well till the precipitate dissolves.
4. Titrate the liberated I_2 with 0.025 M $Na_2S_2O_3$ solution.

Chemical Oxygen Demand (COD)

This is a satisfactory method for determining the organic load of a water body, which is preferable to the biochemical oxygen demand (BOD) mentioned below. It is a rapidly measurable parameter for stream, and industrial waste studies and control of water treatment plants. The method is based on the chemical oxidation of material in the presence of a catalyst by $Cr_2O_7^{2-}$ in 50% H_2SO_4:

$$3\{CH_2O\} + 16H^+ + Cr_2O_7^{2-} \rightarrow 4Cr^{3+} + 3CO_2 + 11H_2O$$

The amount of unreacted $Cr_2O_7^{2-}$ is then determined by titration with a standard Mohr's Salt solution.

Ag_2SO_4catalyses the oxidation of straight chain aliphatic compounds, aromatic hydrocarbons and pyridine. $HgSO_4$ ties up Cl^- as soluble complex and prevents its interference.

Procedure

1. Take 5–50 ml sample in a conical flask (250 ml) with a ground glass joint. Add 10–20 ml 0.25 N $K_2Cr_2O_7$ in 18 N H_2SO_4, 1g Ag_2SO_4 and 1g $HgSO_4$.
2. Reflux for 6 hrs.
3. Cool and titrate excess $K_2Cr_2O_7$ with 0.1 N Mohr salt solution in 8 N H_2SO_4 using 8–10 drops of ferroin (1.485 g o-phenanthroline + 0.695 g. $FeSO_4 . 7H_2O$ dissolved in 100 ml water) as indicator.

Biochemical Oxygen Demand (BOD)

This is an empirical, semi-quantitative method, based on oxidation of organic matter by suitable micro-organisms during a 5-day period. There is nothing sacrosanct about 5 days but the test originated in England where the maximum stream flow is 5 days.

The degree of microbially mediated O_2 consumption in water is known as the *Biochemical Oxygen Demand.* This parameter is commonly measured by the quantity of O_2 utilized by suitable aquatic micro-organisms during 5-day period.

$$\{CH_2O\} + O_2 \xrightarrow{\text{micro-organisms}} CO_2 + H_2O$$

The selection of micro-organisms (seed) is crucial and the results are obviously not reproducible.

The purpose of seeding is to introduce into the sample a biological population capable of oxidizing the organic matter in the waste water. Where such micro-organisms are already present, i.e., in domestic waste water or surface waters, seedling is unnecessary. But when the sample is deficient in micro-organisms, the dilution water needs seedling. The standard seed material is settled domestic waste water which has been stored for 24 to 36 hrs. at 20°C.

Procedure

(a) *Preparation of dilution water*—Take desired volume of distilled water in a suitable bottle and add 1 ml of phosphate buffer (pH 7.2), 1 ml of $MgSO_4$ (22.5 g/l), 1 ml of Cl_2 (27.5 g/l) and 1 ml of $FeCl_3$ (0.25 g/l).

(b) *Dilution technique*—The dilution depends on the nature of the sample: 0.1–1% for industrial wastes: 1–5% for raw and settled sewage; 25–100% for polluted river water and 5–25% for oxidized effluents.

(c) *Procedure*—Aerate the water sample thoroughly by bubbling air through a diffusion tube into the sample for 5 min. or until the DO is 7 ppm. Make a measured dilution with dilution water if BOD is greater than DO level. Seed with a little diluted domestic waste water (1.2 ml/l). Measure DO on a suitable aliquot (D_1).

Fill a screw-topped incubation bottle (250–300 ml) to the brim with the remaining diluted sample. Seal the bottle and incubate in the dark for 5 days at 20°C. Measure DO on an aliquot of the sample (D_2).

$$\text{mg/l BOD} = \frac{(D_1 - D_2) - (B_1 - B_2)f}{P}$$

where D_1 = DO of diluted sample; D_2 = DO of diluted sample after incubation; B_1, B_2 are DO of dilution water containing seed before and after incubation; f = ratio of seed in sample to seed in control and p = decimal fraction of sample used.

BOD of 80 mg/l means that biodegradation of organic matter in a litre of sample consumes 80 mg. of O_2.

Total Organic Carbon (TOC)

This is the best method for the measurement of total organic matter in water samples. As it is an instrumental method, it is superior to the COD method and is quite popular. The oxidizing agent is $K_2S_2O_8$ and the oxidation is promoted by the free radical $HO^\bullet$ generated from ultraviolet light.

$$\text{Organics} + HO^\bullet \xrightarrow{K_2S_2O_8} CO_2\uparrow + H_2O$$

A sketch diagram of a TOC analyzer employing UV irradiation is shown in Fig. 5.2. To the sample $K_2S_2O_8$ and H_3PO_4 are added. H_3PO_4 is sparged with air or N_2 to expel CO_2 formed from HCO_3 and CO_3^{2-} in solution. After sparging, the sample is pumped to a chamber containing a UV lamp emitting radiation at 184.9 nm. The $HO^\bullet$ formed by UV irradiation is the active species which brings about the rapid oxidiation of dissolved organic compounds. At the end of oxidation, CO_2 is sparged from the system and measured with a gas chromatographic detector or by absorption in ultrapure water followed by conductivity measurement.

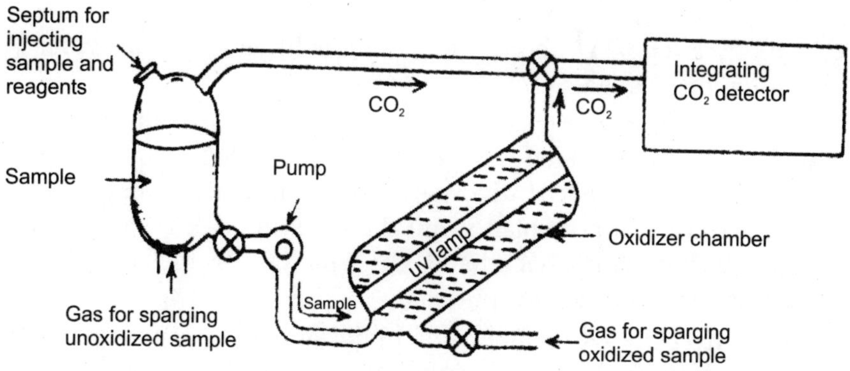

Fig. 5.2. TOC analyzer employing UV-promoted sample oxidation

5.3 WATER POLLUTION CONTROL AND WATER MANAGEMENT—WATER RECYCLING

Clean water is essential for healthy environment to support life systems on this planet. The task of delicately balancing the ratio of available and exploitable water resources and sustaining their quality is most important for India as rainfall distribution is confined to 3–4 months in a year. Moreover, man-made global and local climatic distortions due to global warming (see Chapter 4), deforestation, loss of top soil, etc. have adverse effect on the monsoon pattern in India.

India is blessed with good rainfall (average 200 cm in a year) but 70 per cent of it is wasted. The country faces recurring problems of floods, and droughts and highly polluted water resources. It is necessary to do rain harvesting, i.e., build large tanks and reservoirs all over the country to store rain water, flood water and excess water from the Ganga, Brahmaputra and other rivers. The rivers, the lifelines of our culture and economy, are dying because of severe pollution. This water pollution abatement and resource management should be at the top of our national agenda.

The river water pollution has three dimensions—agricultural run-off, industrial effluents and domestic sewage. A typical example is that of the Ganga pollution. The total quantity of fresh water used in the Ganga basin is 150 billion cubic metres out of which 26 per cent is discharged as waste water. Agricultural run-off is 27 billion cubic metre compared to 1530 million cubic metre from industrial and domestic sewage.

Ganga Action Plan

This best illustrates the strategy of water pollution control and management. In order to rejuvenate (i.e., bring back to life) Ganga, the first step is to stop dumping of all sorts of garbage to the river and at the same time to promote public awareness on this issue. The sources of such garbage are domestic sewage, industrial sewage and agricultural run-off. Every day about 1200 million litres of waste water (mainly from West Bengal) enter the Ganga.

Industrial sewage has to be treated at the plant site through treatment plants and then the treated effluent discharged into the river or recycled for reuse. Domestic sewage in rural and urban areas should be diverted into separate diversion canals and collected in lagoons, subjected to treatment plants and the resulting clean water used for irrigation in agricultural fields. The biogas (methane, CH_4) emitted from the treatment plants can be used for domestic fuel in industrial areas.

With this objective, the Government of India launched Ganga Action Plan in 1986 and entrusted the job to the State Governments concerned (UP, Bihar and West Bengal) for implementation. The Action Plan has the following agenda:

(i) To divert sewage from the small and large towns on both embankments of the Ganga into separate diversion canal, accumulate it in big lagoons, subject it to sewage treatment plant and use the clean sewage water for agriculture. The emitted biogas (methane CH_4) from the plant is utilised as domestic fuel through piped gas supply in adjoining towns.

(ii) To construct low cost sanitary latrines (about 30,000) so that people don't commit nuisance on the river embankments.

(iii) To construct the Ganga embankment ghats and construct electric furnaces at selected sites for funeral rites so that the Ganga is saved from dead bodies.

(iv) To make it mandatory for all industries to install treatment plants in their premises for treatment of their solid and liquid effluent as well as gaseous emission.

For the success of the Ganga Actions Plan, it is essential that the public must be involved—at least they must be made aware of the critical condition of Ganga and need for recovery. For this purpose public awareness campaign deserves top priority. It must be noted that Ganga has three major problems—pollution, erosion and siltation. 70 per cent of pollution is generated from towns (class I 27, class II 23 and class III 48) on both sides of the river, 20 per cent from factories and 10 per cent from agricultural lands. The Ganga Action Plan, however, could not be properly implemented. It required active government-people interaction and participation and furthermore, continuous monitoring of the Projects by experts on a long-term basis. The Ganga Action Plan should be executed as a matter of 20 year plan instead of one or two 5 year plans only.

Desalinisation

Only 0.69 per cent of total water resource is available as fresh water for our consumption. But this resource is shrinking due to reckless water pollution so that in future we may lose it. In that case we have to turn to the vast resource, i.e., sea water (97 per cent), which, at present, is unfit for our consumption due to high salt content. It is possible to utilise sea water with appropriate technology after removal of salts and thus solve water crisis. The technology of removing salts from sea water is known as *desalinisation.* Israel and Arab countries have started applying this technology for solving their chronic problem of water scarcity.

There are several methods for desalinisation of sea water. The most economical method is to use solar energy for evaporation of sea water and separation of salts. For this purpose metal tanks with low height are used for collecting water. The tanks are painted black on the outside walls so that they become hot in the sun while the tanks are covered on the top with transparent plastic sheets. The water vapour from evaporation of sea water by the sun collects on the inside wall of the plastic sheet—on condensation, this water vapour gives pure water. At the end of the operation, salts are recovered from the tanks. Recently Israel is getting each year 30 million cubic metres of salt-free sea water.

Electrodyalysis and Membrane Technology are being used in Japan to obtain salt-free sea water and salts separately. Though these techniques are costly, at present, but with enhanced technology, it may be possible in future to run these on commercial basis.

6 LAND POLLUTION

6.1 LITHOSPHERE–COMPOSITION

Lithosphere: It is made of the mantle of rocks. It includes the soil which covers the rock's crust in many places. Rocks are subjected to continuous weathering forces—rain, wind, chemical and biological—and suffer disintegration. The resulting primitive soil is suitable for the growth of plants—after death and decay, plant debris returns to soil. The mineral component of soil comes from the parent rocks by weathering processes while the organic component is due to plant biomass as well as populations of bacteria, fungi and insects (earthworms). A typical soil, suitable for agriculture, contains about 5 per cent organic matter and 95 per cent inorganic matter. Soil plays an important role as it produces food for man and animals. Good soil and good agriculture are valuable assets for a nation.

In general, soil has a loose structure consisting of solid mineral, organic matter and air space. It shows broadly three zones as its depth increases. The top layer, up to several inches thick, is known as top soil which is an index of soil quality. This is the layer of maximum biological productivity and it contains bulk of organic matter. Hence, it is very important for vegetation cover and agricultural crops. Reckless deforestation causes loss of top soil which means loss of agricultural production. The underlying layer is the subsoil which receives organic matter, salts and clay particles leached from the top soil. The third layer (zone) consists of weathered parent rocks from which soil was formed. Plants draw water and nutrients from soil—they transport water into the plant body (roots and leaves) and discharge excess water into the atmosphere through leaves through the process of transpiration.

The various land forms of the lithosphere are (i) mountain, (ii) plateau and (iii) plain. Mountains provide natural frontier, shelter for tribals, important flora and fauna. They contain forests which are important natural resources. Plateaus are rich in forest and mineral wealth while the plains account for maximum world population because of the convenience for cultivation, communication, transport and industrial growth.

6.2 LAND POLLUTION

Lithosphere with humus cover is known as soil. It is a limited natural resource but most important resource which supports life and contributes to the development of the country. Indeed the soil quality of a nation is its most precious material heritage.

The top soil provides water and all the nutrients required by plants for their growth. The survival of man depends upon agricultural crops, grassland and trees which grow on soil and supply food, fiber, timber, fuel and other products. But it faces threat of erosion from man-made activities. It takes 500–1000 years to build up one inch of top soil while loss of soil takes only a few years. Deforestation alone accounts for loss of 6000 tons of top soil each year.

The land pollution sources are:

(i) indiscriminate solid waste disposal on land;

(ii) random discharge of industrial waste effluents on land and water courses;

(iii) excessive use of pesticides, herbicides and chemical fertilisers, in agriculture;

(iv) excessive irrigation;

(v) open defecation on land by animals and man particularly in rural areas and city slums;

(vi) radioactive materials from radioactive fall-out arising nuclear explosion tests, reactor accidents, etc.

Dumping of solid wastes on land is quite harmful since heavy metals and toxic elements present in the wastes contaminate the soil and ultimately seep into underground water after rainfall. It is estimated that about 15 million tons of solid wastes are generated every year by more than 110 million urban citizens in India.

In general, in rural areas poor illiterate people have the nasty practice of open defecation in absence of sanitary toilets. Both human and animal excreta contain many pollutants which degrade not only soil but also nearby ponds and canals where they mix after being washed by rain water.

Many industries generate huge volume of liquid effluents and solid wastes, which are mostly thrown on land and in water courses (streams and rivers). These effluents and wastes contain phosphorus, chromium, lead, zinc, nickel, cyanides, etc. which affect the quality of soil and render it injurious for plants and animals and ultimately man.

Agriculture run-off (wash water from agricultural fields) contains lot of pesticides, fertilisers, herbicides which find their way into water courses causing water pollution in addition to soil pollution. Excessive irrigation increases the salt content of soil and makes it saline, thereby decreasing soil fertility.

Each nuclear test is accompanied by radioactive fall-out which spreads radioactive contaminants in nearby areas over land and water bodies. Moreover, radioactive wastes from atomic power plants pose long-term health hazards after they contaminate land and water bodies.

Control of Land Pollution

Solid waste disposal methods are actually the methods for control of soil pollution. Non-hazardous wastes are managed by two techniques (i) Landfill and (ii) Incineration. Miscellaneous waste materials—food wastes, vegetable and animal wastes, paper, wood, rubber, leather, plastics, tin cans, crockery, glass, etc. are deposited in a regulated manner in general pit in backyards of houses and municipal areas. These are then covered with 15 cm thick soil. The wastes may be burnt from time to time. Another option for municipal wastes is to use biological process: organic wastes are allowed to be decomposed into humus-like substances after removal of iron and metallic particles. In presence of stream of air, aerobic micro-organisms ensure rapid decomposition of the wastes. Such compost plans are in operation in several parts of the country e.g., Ahmedabad, Mumbai, Chennai, Delhi, Pune, Hyderabad and Kolkata.

The second method viz., Incineration is the preferred technique for waste management particularly in developed countries. It reduces the bulk of waste by 90 per cent at 900–1000°. This is an environment-friendly technique.

Solid Waste Disposal

The methods of waste disposal depend on the nature of waste. Most solid wastes are dumped on land as soil heaps or as landfill to quarries or mine shafts or as dumps consisting or a wide range of materials. Besides these some wastes are dumped into the seas. The various modes of waste disposal are illustrated in Fig. 6.1.

In general, in rural areas poor illiterate people have the nasty practice of open defecation in absence of sanitary toilets. Both human and animal excreta contain many pollutants which degrade not only soil but also nearby ponds and canals where they mix after being washed by rain water.

Many industries generate huge volume of liquid effluents and solid wastes, which are mostly thrown on land and in water courses (streams and rivers). These effluents and wastes contain phosphorus, chromium, lead, zinc, nickel, cyanides, etc. which affect the quality of soil and render it injurious for plants and animals and ultimately man.

Agriculture run-off (wash water from agricultural fields) contains lot of pesticides, fertilizers, herbicides which find their way into water courses causing water pollution in addition to soil pollution. Excessive irrigation increases the salt content of soil and makes it saline, thereby decreasing soil fertility.

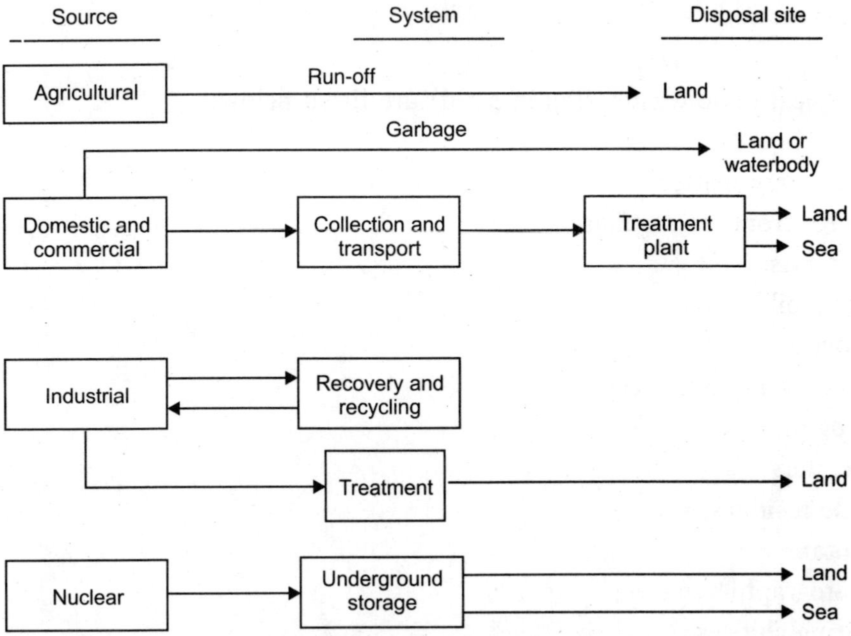

Fig. 6.1: Modes of waste disposal

Each nuclear test is accompanied by radioactive fall-out which spreads radioactive contaminants in nearby areas over land and water bodies. Moreover, radioactive wastes from atomic power plants pose long-term health hazards after they contaminate land and water bodies.

Waste Treatment

Waste water treatment has already been discussed in Chapter 5. Primary and secondary treatment of domestic waste water (municipal waste water) has been discussed. Industrial waste water treatment has also been dealt with.

Disposal of gaseous wastes and their control has been illustrated in Chapter 4. In this chapter only solid waste disposal will be discussed.

Waste Classification and Disposal

Wastes can be broadly classified as domestic, trade and industrial. Domestic waste includes garbage from houses and other residential premises. Industrial waste originates from all mineral, manufacturing and processing establishments.

The break-up of annual solid waste production is given below:

Domestic and trade	8.5 %
Other industries	5.2 %
Thermal power stations	7.3 %

Construction	2%
Mining and quarrying	67%

Some typical toxic wastes (liquid/solid) are listed below:

Tarry liquids;
Solid tarry matter;
Sludge from tar distillation;
Acid tars;
Waste oil;
Lubricants;
Water-kerosene mixtures;
Waste paint;
Lacquers;
Pesticide residues;
Mercaptans;
Photographic wastes;
Plating sludges;
Leaded petrol sludges;
Sludges containing copper, zinc, cadmium;
nickel, arsenic, etc.
Cyanide wastes;
Sulphides, fluorides;
Alkaloid wastes;
Aromatic hydrocarbons;
Chlorophenols;
β-naphthylamine;
Noxious organic solvents

Detoxification of Toxic Wastes

The toxic wastes are converted into less hazardous products by treatment with acids; cyanide containing wastes are decomposed by interaction with oxygen to form carbon dioxide and nitrogen. The decomposition can also be carried out biologically by means of suitable micro-organisms (bacteria). Sludge from petroleum refineries may be spread on the soil and left to decay into harmless products.

Incineration of toxic wastes is another method for their disposal. However, incinerators need pollution control devices and careful monitoring to make sure that they do not release toxic by-products into the environment.

Land and Ocean Disposal

Radioactive wastes from nuclear power stations are generally fused in glass containers and lowered to the ocean floor. In USA such wastes are sealed in metal drums and buried underground at great depths. But they may leak or be damaged by earthquake and release the wastes into ground water.

Hazardous wastes dumped into soil/ditch have chances of leaking to the ground. A typical case history is that of the ***Love Canal*** in Niagara Falls, New York, USA. In 1930–53 the canal ditch was the dumpsite for hazardous chemical wastes and municipal wastes. In 1953 the ditch was filled up; it was covered with clay and sold to the city Board of Education which built an elementary school. Some houses were also built. Soon the residents of these houses (300 families) and the school authorities complained of foul odour and illnesses. In 1978 it was found that some 25 toxic organic compounds known as carcinogens, leaked into the basements in the area and dispersed into air. As a result of these findings, the State of New York declared emergency in the area and transferred all the families and the school from the site.

Non-hazardous Waste Management

Two techniques are available—(i) Landfill and (ii) Incineration.

Sanitary Landfill

Miscellaneous refuse materials are those from household, hotels, stores, markets, restaurants, etc. (e.g., food wastes, vegetable and animals wastes, paper, cardboard, wood, boxes, rubber, leather, plastics, tin cans, crockery glass, metals, etc.), ashes (from fires used for cooking, heating buildings, etc.) dead animals, industries and agricultural fields, etc. In the developed countries, e.g., USA it is a common practice for each household to burn the bulk of waste in a backyard incinerator. Mainly food scraps, bottles and combustible articles are contained in packets for collection by municipality. The percentage of paper, rubber, leather, plastics, metals and glass increases considerably with increasing industrialisation in developed countries compared to developing countries because of wide application of these materials in their daily lives.

Most of the solid waste is dumped on land in heaps in uncontrolled manner in developing countries. Some waste is used for landfilling in abandoned quarries or mines. The developed countries prefer the second method, viz. incineration (see next section). Industrial wastes are treated in treatment plants and valuable materials recycled. In other cases, the volume of the waste is reduced by pulverisation (33 per cent) or by incineration.

For sanitary landfill, the following principles should be followed:

(i) solid wastes should be deposited in a regulated manner, preferably in gravel pit;

(ii) solid wastes should be spread in thin layers with ground cover of at least 15 cm;

(iii) all factors likely to contribute to water pollution should be eliminated;

(iv) the wastes should not be burnt openly.

In the man-made gravel pit fine grained soil is useful in containing undesirable gas and water movements outside the landfill area. With increase in urbanisation and expansion of cities beyond the periphery, land suitable for landfill becomes more and more scarce. In such cases planned sanitary landfill, backed by modern solid waste management can provide the community with better environmental management.

Composting and Municipal Waste Composting Projects

This is biological process where fresh organic wastes are allowed to be decomposed into humus-like substances. The process is conducted by a complete automatic system which consists of several steps: (1) The crude refuse is dumped into container or to a belt conveyor. (2) Iron or metallic particles are removed by a magnetic separator. (3) The material is then transferred in a wet condition to a rotatory cylinder, analogous to a rotatory drier. The cylinder rotates slowly on large tyres and the wastes move from one end to the other. They are thoroughly mixed and pulverised by abrasion. Air is introduced at low pressure throughout the length of the cylinder. Here aerobic micro-organisms ensure rapid decomposition of the wastes under aerobic conditions.

In several parts of India compost plants are in operations, e.g., in Ahmedabad, Kolkata, Mumbai, Chennai, Delhi, Pune and Hyderabad. The minimum plan capacity is 100 tons per day. The Government encourages feeding of compost plant by municipal wastes as raw materials. Compost has been used by Indian Agricultural Research Institute (IARI), New Delhi to develop strains of blue-green algae for application to rice fields with a view to nitrogen fixation. An appropriate technology is being developed by IARI to produce blue-green algae-coated granulated compost.

Incineration

This is the preferred technique for waste management, particularly in the developed countries. It reduces the waste volume by 90 per cent at 900–1000°C. Incineration offers environment-friendly technique—free from corrosion, emission of offensive odours and also free from bacteria and wet organic matter which gives off foul odours and gases. The waste heat from incineration can be utilised for supplementing electricity generation for domestic heating, etc. The only drawback is that the technique is costly at present requiring expensive equipment.

Two types of incinerators are used for unsorted wastes. The batch type plant is manually stoked and has a relatively small rated capacity. The operation is intermittent and lacks uniform burning temperature. This leads to incomplete combustion and yields an unstable residue. These units are not suitable for large cities. The continuous feed plant has larger storage bins, automatic feed hoppers and variety of moving gates and ash removal systems. The unit maintains a uniform combustion range, can be fitted with pollution control devices and yields stable residue.

Japan since 1970 and Germany since 1990 developed recycling technology of plastic wastes. The latter was subjected to incineration and gasification—the gaseous products were converted to chemical raw materials (water gas, i.e., carbon monoxide and hydrogen mixture which can be utilised for production of methanol) and the heat generated used for electrical generation units or for domestic heating. The waste volume shrinks to 20 per cent and carbon dioxide released to the atmosphere is minimum so that the technique offers good recycling.

Ocean Dumping

In the developed countries such as USA about 50–75 million tons of waste are dumped every year within 200–300 km of ocean shore. In earlier years waste dumped into the ocean 100 km from the shore did little damage. But over the years industrial wastes and domestic wastes have become more toxic.

Industrial wastes contain numerous toxins and sewage sludge also contains lot of toxic metals. Disease-causing organisms and heavy metals have destroyed many coastal fisheries. Anti-dumping Act (1972) has been violated and many metropolitan cities—New York, Tokyo, etc.—continue to dump sewage into the ocean. It has been the practice in all countries including India to throw radioactive wastes into the oceans and high seas.

Solid Waste Management by Biotechnology

According to biologists, bacteria and fungi are capable of decomposing organic waste and it may be possible to recover resources by this process. Natural micro-organisms can do this job—it is also possible to produce such micro-organisms by genetic engineering. The promising development is the isolation of bacteria which can break down polychlorinated biphenyls (PCBs).

New biodegradable plastics are important step towards solving our solid waste problems in respect of plastic wastes. On exposure to micro-organisms which metabolise glucose, biodegradable polymers break down into short carbon chains that decomposers can metabolise. Photodegradable plastics have been developed, which break down on exposure to sunlight.

❍❍❍

7 NOISE POLLUTION

7.1 NOISE POLLUTION

Noise is part of our environment. With progress in industrialisation, the noise level has been rising continuously. In the 19th century the development of the steam engines, petrol engine and machines in factories resulted in increasingly noisy environment. In the 20th century this was further accelerated by introduction of diesel engine, jet engines, turboprop, high tech machineries, construction site machineries and automobile traffic. Noise has been recognised as one of the dimensions of pollution which brings about degradation of the environment and creates health and communication hazards.

7.2 SOUND AND HUMAN ACOUSTICS

Sound consists of wave motion in an elastic medium such as air, water or solids (e.g., metals, plastics, wood, bricks, etc.). Sound waves travel through the medium from the source to the recipient or listener. The rate of the oscillation of the medium is known as the *frequency* of the sound, the unit being *Hertz* (*Hz*) or *cycles per second*. The frequency is a measure of the pitch of the sound received by the listener. High frequencies mean high pitched sounds which are more irritating to the individual than low frequencies. The second parameter of sound is *sound pressure* which is measured in newtons per sq. metre (N/m^2). The third parameter on sound is its *intensity,* expressed in watts per sq. metre i.e., the quantum of sound energy that flows through unit area of the medium in unit time.

The human ear receives sound waves which set up oscillations in the ear drum (tympanic membrane). These oscillations cause movement of three small bones in the middle ear behind the ear drum. These then pass through the fluid in the inner ear to the auditory nerve and finally transmitted to the brain. The oscillations or sound are identified and interpreted in the brain, which can select sounds into different categories—speech, music, noises, etc.

The sensitivity of the ear varies from person to person. With ageing, people lose hearing power gradually. A young person, 18 year old, with normal hearing, has audio range between 20 Hz and 20,000 Hz. The audio sense is sharpest in the frequency range 2000–8500 Hz.

7.3 NOISE MEASUREMENT UNITS

As mentioned before, sound pressure and sound intensity are two important parameters of noise. The common scientific acoustic unit is the **Decibel (dB).** It is not an absolute physical unit like volt, metre, etc. but is a ratio, expressed in logarithmic scale relative to a reference sound pressure level.

$$1 \textbf{ decibel } (\text{dB}) = 10 \log_{10} \frac{\text{intensity measured}}{\text{reference intensity}}$$

The reference intensity used is the threshold of hearing which means sound which can be first heard at a sound pressure of 2×10^{-5} newton per sq. metre or sound intensity of 10^{-12} watts per sq. metre.

Noise meters have been designed for noise measurement from low to high frequencies, characteristic of human ear capacity. These meters record the dB scale for routine measurement of general noise levels. Refined noise meters have been developed to take care of peak noise levels, duration of noise exposure and quality of noise which are aspects of specified noise situation.

Table 7.1: Sound Measurement (Intensities, Pressures and Decibels) in Air at Room Temperature and Sea Level Pressure

Intensity (WM^{2-})	*Pressure (Nm^{2-})*	*dB*	*Sound source*
100	2,00,000	200	Saturn rocket take-off
1.0	20	120	Boiler shop
10^{-2}	2.0	100	Siren at 5 metres
10^{-4}	0.2	80	Heavy machinery
10^{-6}	0.02	60	Normal conversation at 1 metre
10^{-8}	0.002	40	Public library
10^{-12}	2×10^{-5}	0	Threshold of hearing

7.4 NOISE CLASSIFICATION

There are broadly three categories of noise:

(i) Transport noise, (ii) Occupational noise, and (iii) Neighbourhood noise.

Transport Noise

Transport noise can be further sub-divided into (i) Road traffic noise, (ii) Aircraft and (iii) Rail traffic noise.

Road Traffic Noise

Traffic noise is increasing over the years with increase in the number of road vehicles. Traffic speed is the major cause of noise. The noise volume is enhanced with increase in traffic speed. Modern highways and traffic system encourage higher speeds.

In general, on urban roads there are distinct traffic peaks in the morning and evening (10 A.M. and 6 P.M.) as people travel to and fro workplaces. Heavy diesel-engined trucks are the noisiest vehicles on roads at present. The permissible noise levels for cities are prescribed by the Central Pollution Control Board, India:

Table 7.2: Occupational Noise Level

Areas	*Day*	*Night*
Industrial	75 dB	65 dB
Commercial	65 dB	55 dB
Residential	50 dB	45 dB
Sensitive areas up to 100 m around hospitals, schools	50 dB	40 dB

These limits are, however, violated in all big cities in India, Kolkata being the worst case. The average noise levels in busy streets in Kolkata during rush hours (between 10:30–12:00 hrs. and 18:00–19:30 hrs.) are 90 dB. People live in an environment of noise generated by blasting horns, rumbling tyres and screeching brakes. Awful road accidents contribute to the misery.

Occupational Noise

Industrial workers are exposed to noisy working environment for 48 hours a week (8 hrs. a day for 6 days a week). Some typical occupational noise levels are given below:

Table 7.3: Occupational Noise Level

Industrial source	*Noise level (dB)*
Steel plate riveting	130
Oxygen torch	126
Boilers' shop	120

contd...

Textile loom	112
Circular saw	110
Farm tractor	103
Newspaper press	101
Bench lathe	95
High speed drill	85
Supermarket	60

Millions of workers suffer from progressive hearing damage and become prone to accidents under their working conditions. Their working efficiency is also affected.

Neighbourhood Noise

Loud TV and radio sets, loud cassettes, loudspeakers in public functions, disco music, etc. are sources of neighbourhood noise which disturb and irritate the general public and also harm the patients.

7.5 NOISE POLLUTION HAZARDS

The human ear drum is struck by noise in the form of airborne mechanical energy. While the tolerable conversation level is 65 dB at a distance of 1 metre, 125 dB gives the sensation of pain in the ear and 150 dB might be a killer.

High intensity noise for continuous periods is the major cause for ear damage. If a noise level exceeding 90 dB in the mid-frequency range reaches the ear for more than a few minutes, then the sensitivity of the ear is reduced.

Noise pollution can cause pathological or psychological disorders. High frequencies or ultrasonic sound above the audible range can affect the semi-circular canals of the inner ear and make one suffer from nausea and dizziness. Mid-audible frequencies can lead to resonance in the skull and thereby affect the brain and nervous system. Moderate vibration can also cause pain, numbness and cyanosis (blue colouration) of fingers while severe vibration results damage to bones and joints in the bands with swelling and stiffness.

In industrial and other establishments the general impact of noise pollution is lower efficiency, reduced work rate and higher potential for accidents and injuries.

In residential areas even low frequency noise of 50–60 dB at night disturbs sleep, particularly among the aged people, causing adverse effect on health.

Children, exposed to excessive noise, show signs of behavioural disorder which in later age develop into destructive nature and neurotic disorders in the adult.

Excessive noise is one of the major factors for chronic exhaustion and tension in our daily lives. This may explain why more and more people tend to become addicted to alcohol, tobacco and drugs.

Noise pollution has also impact on travel of migratory birds from winter to tropical climate. Thus the increase of noise pollution in Kolkata and construction of high-rise buildings near Alipur Zoological garden have led to decline in the number of migratory birds from CIS (former USSR) from 15,000 in 1980 to 2000 in 1990.

7.6 PERMISSIBLE NOISE LEVELS

In this age many people work and live in environment where the noise level is not hazardous. But over the years they suffer from progressive hearing loss and psychological hazards. The maximum permissible noise levels are summarised as follows:

Table 7.4: Maximum Permissible Noise Levels

Situation	*Permissible noise, dB*
Road traffic near residential areas	70
Ear protection required	85
Factory work (48 hr. week)	105
Prolonged noise causing permanent damage	100
Threshold of pain (30 secs duration)	120
Maximum for sonic boom	150
Ear drum rupture	180

QUESTIONS (CHAP 5–7)

1. List the water pollutants. How do they cause water pollution?
2. Name the major water quality parameters mentioning their permissible limits.
3. Comment on As-contamination of ground water and As-speciation.
4. What are the causes of land pollution? How can it be controlled?
5. Illustrate with a flow sheet diagram municipal waste treatment of solid waste.
6. What is meant by Decibel of noise?
7. Give a brief account of noise pollution hazards.
8. What are permissible noise levels in residential areas and factories?
9. How will you monitor:
   (a) D.O. (b) COD (c) BOD (d) TOC
   in water samples?
10. Write notes on:
    (a) Sanitary landfill method for waste disposal
    (b) Incineration method for waste disposal
    (c) Municipal waste composting
    (d) Ganga Action Plan
    (e) Water recycling

❍ ❍ ❍

8 Feedback Exercises

The Feedback Exercises are intended to help the students to test their own understanding of the subject matter of this book which in turn will sustain their interest in Environment.

Put a tick mark (√) within the bracket against the correct answer(s). If there are more than one correct answer, both should be marked. Some answers are given at the end indicating the serial number.

1. Environment means
   (a) a beautiful landscape ( )
   (b) industrial production ( )
   (c) sum total of all conditions that effect the life and development of all organisms on earth ( )
   (d) air and water ( )
2. The elements of environment are
   (a) nitrogen, oxygen and carbon dioxide ( )
   (b) nitrogen, oxygen and sulphur dioxide ( )
   (c) oxygen, ozone and carbon dioxide ( )
   (d) atmosphere, hydrosphere and lithosphere ( )
3. Environment has two major components [NEHU, 2000]
   (a) plants and animals ( )
   (b) man and nature ( )
   (c) biotic and abiotic ( )
   (d) physical and chemical ( )
4. Atmosphere has the main constituents
   (a) nitrogen, oxygen and water vapour ( )
   (b) argon, neon and helium ( )
   (c) carbon dioxide, methane ( )
   (d) nitrogen oxides and sulphur oxides ( )

5. Troposphere has altitude range
   (a) 11–50 km ( )
   (b) 0–11 km ( )
   (c) 20–80 km ( )
   (d) 80–200 km ( )
6. Stratosphere contains an important species which protects life on earth
   (a) oxygen ( )
   (b) nitrogen ( )
   (c) ozone ( )
   (d) carbon dioxide ( )
7. Hydrological cycle provides us with
   (a) supply of fresh water ( )
   (b) irregular rainfall ( )
   (c) fresh oxygen ( )
   (d) electricity ( )
8. Hydrosphere denotes
   (a) water ( )
   (b) plants ( )
   (c) micro-organisms ( )
   (d) soil ( )
9. Lithosphere means
   (a) air ( )
   (b) water ( )
   (c) rocks and soil ( )
   (d) organisms ( )
10. A zone consisting of land, water and air which supports life is called
    (a) biosphere ( )
    (b) atmosphere ( )
    (c) lithosphere ( )
    (d) hydrosphere ( )
11. Ecosystem denotes
    (a) seas and oceans ( )
    (b) functional unit in ecology ( )
    (c) forests ( )
    (d) natural cycles ( )
12. The three major living components of an ecosystem are
    (a) producers, phototrophs and consumers ( )
    (b) producers, decomposers and consumers ( )

(c) proteins, carbohydrates and humic acids ( )
(d) micro-consumers, bacteria and fungi ( )

13. For manufacturing their own food, plants are classified as
(a) consumers ( )
(b) producers ( )
(c) decomposers ( )
(d) food web ( )

14. In an ecosystem there is dynamic balance between
(a) man, animals and birds ( )
(b) minerals, water and soil ( )
(c) plants, bacteria and micro-organisms ( )

15. Bacteria and fungi are
(a) decomposer ( )
(b) producers ( )
(c) consumers ( )
(d) none of the above ( )

16. The most important organisms of an ecosystem are
(a) herbivores ( )
(b) carnivores ( )
(c) green plants ( )
(d) protozoa ( )

17. Forests bring about
(a) air pollution control ( )
(b) soil erosion ( )
(c) cyclones ( )
(d) increase in soil fertility ( )

18. Deforestation leads to
(a) heavy rainfall ( )
(b) enhanced soil erosion ( )
(c) balance of ecosystem ( )
(d) none of the above ( )

19. Afforestation is necessary for
(a) soil conservation ( )
(b) soil erosion ( )
(c) weed control ( )
(d) low humidity ( )

20. In the hills forest cover is affected by
(a) overgrazing by cattle ( )
(b) soil erosion by rain ( )
(c) leaching of soil ( )
(d) soil pollution ( )

21. Social forestry is meant for
(a) supplying goods and services to people ( )
(b) providing services to industries ( )
(c) putting pressure on forests ( )
(d) improvement of degraded lands ( )

22. Extinction of species is due to
(a) natural process ( )
(b) man-made activities ( )
(c) animal pollution ( )
(d) industrial pollution ( )

23. How many living organisms on earth are on record?
(a) 1.7 million ( )
(b) 16 million ( )
(c) 30 million ( )
(d) nil ( )

24. Biodiversity means
(a) the living natural resources ( )
(b) micro-organisms only ( )
(c) seas and oceans ( )
(d) lithosphere ( )

25. The energy flow through the ecosystem is (NEHU, 2000)
(a) cyclic ( )
(b) linear and one way ( )
(c) both cyclic and linear ( )
(d) linear and two-way ( )

26. The transfer of food energy takes place in some hierarchical order called
(a) Ecosystem ( )
(b) Trophic levels ( )
(c) Ecological pyramids ( )
(d) Ecological niche ( )

27. Primary producers produce their own food by
(a) converting solar energy into chemical energy ( )
(b) hydroelectricity ( )
(c) osmosis ( )
(d) mechanical energy ( )

28. A food chain is
  (a) a group of organisms which eat the same type of food ( )
  (b) animals eating the plants ( )
  (c) series of plants and animals which are interrelated in the process of eating and being eaten up ( )
  (d) none of the above ( )

29. A food chain always starts with
  (a) photosynthesis ( )
  (b) respiration ( )
  (c) nitrogen fixation ( )
  (d) decay ( )

30. A food web
  (a) increases variety of food at each trophic level ( )
  (b) delicately balances the interrelations among organisms ( )
  (c) increases the quantity of food at each trophic level ( )
  (d) increases variety as well as quantity of food at each trophic level ( )

31. A group of interconnected food chain is called
  (a) complex food chain ( )
  (b) food cycle ( )
  (c) food web ( )
  (d) none of the above ( )

32. The current world population is [NEHU, 2001]
  (a) 4 billion ( )
  (b) 5 billion ( )
  (c) 6 billion ( )
  (d) 8 billion ( )

33. The average number of persons living per sq. km. is known as
  (a) community ( )
  (b) population density ( )
  (c) population growth rate ( )
  (d) population explosion ( )

34. It is estimated that the world population may stabilise around 2050 at [NEHU, 2000]
  (a) 10 billion ( )
  (b) 12 billion ( )
  (c) 14 billion ( )
  (d) 20 billion ( )

35. The largest metropolitan city in India (1991) was
(a) Greater Mumbai ( )
(b) New Delhi ( )
(c) Kolkata ( )
(d) Chennai ( )

36. The number of 1-million plus cities in India in 1991 was
(a) 20 ( )
(b) 23 ( )
(c) 25 ( )
(d) 28 ( )

37. The state with the highest population growth rate between 1981 and 1991 is
(a) U.P. ( )
(b) Bihar ( )
(c) Assam ( )
(d) Nagaland ( )

38. At the present growth rate India's population in 2010 will reach
(a) 1050 million ( )
(b) 1100 million ( )
(c) 1200 million ( )
(d) 1500 million ( )

39. High fertility in India is associated with
(a) early marriage ( )
(b) low literacy level ( )
(c) mother's malnutrition ( )
(d) high infant mortality ( )

40. India has doubling time for population
(a) less than 25 years ( )
(b) 50 years ( )
(c) 60 years ( )
(d) 75 years ( )

41. Population explosion occurs
(a) when natural disaster strikes ( )
(b) when population size exceeds the carrying capacity of a country ( )
(c) under conditions of low fertility ( )
(d) when infant mortality is low ( )

42. Population pressure affects
(a) food and natural resources ( )
(b) has no effect on livestock ( )

(c) has no effect on fuel stock ( )
(d) poverty and pollution ( )

43. Primary air pollutants are
(a) oxygen, nitrogen and water vapour ( )
(b) carbon monoxide, nitrogen oxides, sulphur oxides and hydrocarbons ( )
(c) carbon dioxide, sulphur dioxide and ozone ( )
(d) halogens and methane ( )

44. Automobile exhausts consist of
(a) hydrocarbons, carbon monoxide and nitric oxide ( )
(b) lead vapour ( )
(c) sulphur dioxide ( )
(d) carbon dioxide ( )

45. Auto emissions are controlled by
(a) using oxygen stream in the exhaust pipes ( )
(b) using catalytic converters in the engines ( )
(c) substituting petrol by alcohol ( )
(d) using less fuel and more air ( )

46. Acid rain originates from
(a) acetic and phosphoric acids ( )
(b) hydrogen chloride, nitric and sulphuric acids ( )
(c) sulphuric acid ( )
(d) leather industries ( )

47. The source of acid rains are
(a) steel plants ( )
(b) nuclear reactors ( )
(c) thermal power plants ( )
(d) coke oven plants ( )

48. Fine particles (0.01–1.0 μ size) cause
(a) typhoid ( )
(b) dysentery and diarrhoea ( )
(c) asthma, bronchitis, tuberculosis ( )
(d) cholera ( )

49. The source of soot particle in air is
(a) burning of paper ( )
(b) fuel combustion ( )
(c) evaporation of water from seas ( )
(d) coke oven plants ( )

50. Particulate emissions from gas streams can be removed by
   (a) lightning discharge in air ( )
   (b) electrostatic precipitator ( )
   (c) air jet ( )
   (d) acid spray ( )

51. Meteorology of the earth is affected by
   (a) natural disasters ( )
   (b) human activities ( )
   (c) earth's rotation ( )
   (d) tidal waves ( )

52. Fresh water available for use is
   (a) less than 10 per cent of total water resources ( )
   (b) less than 5 per cent of total water resources ( )
   (c) less than 1 per cent of water resources ( )
   (d) more than 2 per cent of total water resources ( )

53. Ocean water, vast water resource (97 per cent), is
   (a) good for agriculture ( )
   (b) unfit for human consumption ( )
   (c) useful for coastal vegetation ( )
   (d) useful for inland fisheries ( )

54. Ground water, compared to sea water and lake water, has
   (a) same minerals salts ( )
   (b) less mineral salts, nitrate and bicarbonate ( )
   (c) more mineral salts, nitrate, carbonate ( )
   (d) is fit for our use ( )

55. The symptoms of polluted water are
   (a) no change in physical appearance ( )
   (b) no external matter on the surface ( )
   (c) foul smell, bad taste, dead fish floating on the surface ( )
   (d) none of the above ( )

56. Arsenic contamination of ground water arises from
   (a) excessive withdrawal by shallow tube wells ( )
   (b) leakage of surface water ( )
   (c) leaching of arsenopyrite mineral into ground water ( )
   (d) dumping of industrial wastes on ground ( )

57. Arsenic poisoning symptoms are
    (a) diarrhoea, dysenetery, etc. ( )
    (b) loss of hair, brittle nails, gangrene and cancer ( )
    (c) pneumonia, typhoid, etc. ( )
    (d) chest congestion ( )

58. The Ganga river is heavily polluted at
    (a) Rishikesh and Haridwar ( )
    (b) Kanpur, Varanasi and Kolkata ( )
    (c) Patna and Allahabad ( )
    (d) Hyderabad ( )

59. The Ganga pollution is due to dumping of
    (a) domestic and industrial sewage ( )
    (b) food wastes ( )
    (c) waste from forests ( )
    (d) industrial gases ( )

60. For drinking water supply, water treatment requires
    (a) filtration through sand bed ( )
    (b) chlorination for killing viruses ( )
    (c) sedimentation ( )
    (d) aeration ( )

61. The important water quality parameters are
    (a) pH, dissolved oxygen, chemical oxygen demand, chloride, arsenic ( )
    (b) calcium and magnesium ( )
    (c) lead, zinc and mecury ( )
    (d) all of the above ( )

62. Desalinisation of sea water means
    (a) removal of salts from sea water ( )
    (b) dumping of oil into sea water ( )
    (c) removal of suspended matter from sea water ( )
    (d) aeration of sea water ( )

63. Toxic wastes are handled by conversion into less hazardous product(s) by treatment with
    (a) acids ( )
    (b) alkalies ( )
    (c) aeration ( )
    (d) chlorination ( )

64. Non-hazardous wastes are managed by
(a) dumping on roads ( )
(b) sanitary landfill ( )
(c) composting process ( )
(d) accumulation at the site ( )

65. Radioactive substances emit radiations
(a) alpha and beta ( )
(b) alpha, beta and gamma ( )
(c) cosmic rays ( )
(d) ultraviolet ( )

66. Radioactive radiations have
(a) ionising power ( )
(b) penetrating power in air ( )
(c) no effect on matter ( )
(d) no effect on health ( )

67. The radioactive disintegration series have the same end product but different isotopes
(a) copper ( )
(b) iron ( )
(c) lead ( )
(d) bismuth ( )

68. Atom bomb explosions destroyed
(a) Moscow ( )
(b) Hiroshima and Nagasaki ( )
(c) Bangkok ( )
(d) Manila ( )

69. Excessive radiation exposure leads to
(a) cancer and death ( )
(b) typhoid ( )
(c) bronchitis ( )
(d) genetic damage ( )

70. Noise pollution becomes hazardous above
(a) 90 decibel ( )
(b) 100 decibel ( )
(c) 120 decibel ( )
(d) 150 decibel ( )

71. Noise levels in residential areas should not exceed
(a) 100 decibels ( )
(b) 150 decibels ( )

(c) 70 decibels ( )
(d) 85 decibels ( )

72. Air-borne diseases are
(a) asthma, bronchitis, pneumonia, etc. ( )
(b) cholera and dysentery ( )
(c) malaria ( )
(d) jaundice ( )

73. Renewable resources are
(a) coal, natural gas ( )
(b) forests, plants, water ( )
(c) agricultural crops ( )
(d) all of the above ( )

74. Non-renewable resources are
(a) petroleum and natural gas ( )
(b) solar energy ( )
(c) forests ( )
(d) rains and lightning ( )

75. Minerals mined to the maximum extent are
(a) coal, petroleum, iron ore ( )
(b) copper, zinc ( )
(c) chromium, limestones ( )
(d) sodium chloride, magnesium chloride ( )

76. Green Revolution resulted from
(a) high-input high-output technique of agriculture ( )
(b) use of fertilisers in agriculture ( )
(c) excessive afforestation ( )
(d) production of green vegetables ( )

77. Green Revolution concerned
(a) production of wheat ( )
(b) production of maize ( )
(c) production of fish ( )
(d) production of cotton ( )

78. Fossil fuel means
(a) paper and wood ( )
(b) hydroelectricity ( )

(c) coal, petroleum and diesel ( )
(d) food waste ( )

79. Conventional energy resources are
(a) fossil fuel, wood, natural gas ( )
(b) tidal wave energy ( )
(c) storms and lightning ( )
(d) nuclear energy ( )

80. The most abundant energy resource is
(a) diesel ( )
(b) coal ( )
(c) solar energy ( )
(d) electricity ( )

81. The major renewable natural energy resource is
(a) nuclear fuel ( )
(b) wood ( )
(c) coal ( )
(d) dry batteries ( )

82. Sunlight may be converted into electricity through
(a) galvanic cell ( )
(b) carbon electrodes ( )
(c) photovoltaic cell ( )
(d) glass blocks ( )

83. The clean pollution free energy resource is
(a) coal ( )
(b) solar energy ( )
(c) petroleum ( )
(d) nuclear power ( )

84. Sustainable development means
(a) development without considering its impact on environment ( )
(b) balance between development and environment with due emphasis on conservation of resources ( )
(c) development at the cost of environment ( )
(d) none of the above ( )

Answers to the Multiple Choice Questions

1. (c)	2. (d)	3. (c)	4. (a)	5. (b)	6. (c)	7. (a)	8. (a)
9. (c)	10. (a)	11. (b)	12. (b)	13. (b)	14. (c)	15. (a)	16. (c)
17. (a)	18. (b)	19. (a)	20. (b)	21. (a)	22. (b)	23. (c)	24. (a)
25. (b)	26. (b)	27. (a)	28. (c)	29. (a)	30. (d)	31. (c)	32. (c)
33. (b)	34. (c)	35. (a)	36. (b)	37. (d)	38. (c)	39. (a,b)	40. (a)
41. (b)	42. (a,d)	43. (b)	44. (a)	45. (b)	46. (b)	47. (c)	48. (c)
49. (b)	50. (b)	51. (b)	52. (c)	53. (b)	54. (b,d)	55. (c)	56. (a)
57. (b)	58. (b)	59. (a)	60. (b)	61. (a)	62. (a)	63. (a)	64. (c)
65. (b)	66. (b)	67. (c)	68. (b)	69. (a)	70. (a)	71. (c)	72. (a)
73. (b)	74. (a)	75. (a)	76. (a)	77. (a)	78. (c)	79. (a)	80. (c)
81. (b)	82. (c)	83. (b)	84. (b)				

Sample Question Papers

SAMPLE PAPER– I

The figures in the margin indicate full marks

1. Answer **any twelve** questions of the following giving tick mark (✓) mark in appropriate box:

2 × 12

1.1 Non-living components of ecosystem are
(a) Biotic ☐ (b) Abiotic ☐
(c) Free living ☐ (d) None ☐

1.2 Acid rain is caused by oxides of
(a) SO_2 ☐ (b) NO_2 ☐
(c) Both ☐

1.3 Converting solid waste into new products by using resources contained in the discarded material is
(a) Waste management ☐ (b) Reuse ☐
(c) Recycling ☐

1.4 Which amongst the following is decomposer in ecosystem?
(a) Green plant ☐ (b) Animal ☐
(c) Bacteria ☐ (d) (a) and (b) ☐

1.5 Acidity in rain is measured by
(a) Barometer ☐ (b) Hygrometer ☐
(c) Ammeter ☐ (d) pH meter ☐

1.6 The process by which anaerobic decomposition of organic matter by bacteria gives out a bad smell is

(a) Decomposition ☐ (b) Fermentation ☐
(c) Putrefaction ☐

1.7 Profuse growth of aquatic vegetation that often changes the colour of water and reduces the dissolved oxygen content is called

(a) Algal bloom ☐ (b) Eutrophication ☐
(c) Fermentation ☐ (d) (a) and (b)

1.8 Montreal Protocol signed in September 1987 aims at the control of

(a) Use of ozone depleting substances ☐
(b) Use of greenhouse gases ☐
(c) Use of chemical pesticides ☐

1.9 Stone cancer or stone leprosy is caused by

(a) UV rays ☐ (b) Greenhouse effect ☐
(c) Acid rain ☐ (d) Ozone depletion ☐

1.10 Tree hugging movement is

(a) Chipko Andolan ☐ (b) Silent Valley movement ☐
(c) Green movement ☐

1.11 Rio declaration refers to

(a) Earth Summit in June 1992 ☐ (b) Ramsar conference ☐
(c) Stockholm conference 1972 ☐

1.12 World Environment Day is observed on

(a) April 22 ☐ (b) June 22 ☐
(c) June 5 ☐

1.13 Ozone layer depletion occurs in

(a) Stratosphere ☐ (b) Troposphere ☐
(c) Ionosphere ☐

1.14 Global warming is caused by

(a) Acid rain ☐ (b) Ozone depletion ☐
(c) Greenhouse gases ☐

1.15 The source of oxygen in atmosphere is due to

(a) Respiration ☐ (b) Excretion ☐
(c) Photosynthesis ☐

1.16 The scale that measures the magnitude of earthquake is
(a) Kelvin scale ☐ (b) Fahrenheit scale ☐
(c) Richter scale ☐

1.17 At the time of formation of earth, free oxygen in atmosphere was
(a) Available ☐ (b) Not available ☐

1.18 CFC is
(a) Chlorofluorocarbon ☐ (b) Centre for fuel control ☐
(c) Carcinogenic fluoride compound ☐

1.19 Which amongst the following is primary consumer?
(a) Bacteria ☐ (b) Tiger ☐
(c) Goat ☐ (d) Vulture ☐

1.20 Of the following, water borne disease is
(a) Small pox ☐ (b) Cholera ☐
(c) Diabetes ☐

2. Answer **only ten** questions (within 2–3 lines) 2 × 10

2.1 Define atmosphere.
2.2 What are the renewable and non-renewable energy sources? Give one example for each.
2.3 Name the principal gases of atmosphere.
2.4 State the importance of mangrove forest.
2.5 Define a producer (in ecology) giving an example.
2.6 What is wild life sanctuary?
2.7 Describe two major air pollutants emitted by automobiles.
2.8 What is BOD?
2.9 What is biomass?
2.10 What is wetland? Name two important wetlands in India.
2.11 Name three sources of odour pollution.
2.12 What is the permissible noise level at night as prescribed by the Central Pollution Control Board in (i) residential areas and (ii) in silent zones?
2.13 Define biogeochemical cycle.
2.14 State the impact of land degradation on agriculture.
2.15 What is carrying capacity of environment?
2.16 What is Green Bench?
2.17 What is Agenda 21?

2.18 Explain what is biomagnification.
2.19 Explain what is composting.
2.20 Name two Project Tiger Reserves in West Bengal.

3. Answer **any one** question (within 120 words) 6 × 1

3.1 Explain what is biodiversity and the necessity for its preservation.
3.2 Name two non-conventional energy sources and mention their possible applications in Indian context.
3.3 Give an outline of natural and man-made disaster.
3.4 What are the measures to be taken to prevent water pollution-related diseases?
3.5 What is nitrogen cycle? State the role of bacteria in nitrogen cycle.
3.6 State the importance of wetland.

SAMPLE PAPER–II

The figures in the margin indicate full marks

1. Answer **only twelve** questions putting tick mark (✓) marks in appropriate box: 2× 12

1.1 The earliest traces of life on earth have been found in the rocks of
(a) 3.5 trillion years old ☐ (b) 3.5 billion years old ☐
(c) 3.5 million years old ☐

1.2 Richter scale is used for measuring
(a) Velocity of light ☐ (b) Intensity of sound ☐
(c) Amplitude of seismic waves ☐

1.3 Bhopal gas disaster occurred in
(a) 1964 ☐ (b) 1974 ☐ (c) 1984 ☐

1.4 In 1986, nuclear plant disaster took place in
(a) London ☐ (b) Chernobyl ☐ (c) Three miles ☐

1.5 First UN conference on Environment was held in
(a) Johannesburg ☐ (b) Stockholm ☐ (c) Montreal ☐

1.6 Which one of the following gases is most abundant in atmosphere?
(a) Methane ☐ (b) Nitrogen ☐ (c) CFC ☐

1.7 The interiormost layer of the earth is known as
(a) Core ☐ (b) Crust ☐ (c) Mantle ☐

1.8 The landmass of the earth is known as
(a) Biosphere ☐ (b) Lithosphere ☐ (c) Stratosphere ☐

1.9 Mangroves are found in
(a) Puskar lake ☐ (b) Sundarbans ☐ (c) Loktak lake ☐

1.10 Which one of the following is a pollution-free renewable resource?
(a) Fossil fuel ☐ (b) Solar energy ☐ (c) Wood ☐

1.11 What is the name of the supercontinent which broke into the present day continents?
(a) Pampas ☐ (b) Palaearctic ☐ (c) Pangaea ☐

1.12 Ozone hole is present in
(a) Biosphere ☐ (b) Troposphere ☐ (c) Stratosphere ☐

1.13 Which one of the following is an autotroph?
(a) Green plant ☐ (b) Fungus ☐ (c) Fish ☐

1.14 Human malaria is transmitted by
(a) *Anopheles* male ☐ (b) *Anopheles* female ☐
(c) *Culex* male ☐

1.15 The safe intensity level of sound is
(a) 85 decibel ☐ (b) 75 decibel ☐ (c) 65 decibel ☐

1.16 The Silent Valley is in
(a) Andhra Pradesh ☐
(b) Himachal Pradesh ☐
(c) Kerala ☐

1.17 Which one of the following is not a gaseous cycle?
(a) Oxygen cycle ☐
(b) Phosphorus cycle ☐
(c) Nitrogen cycle ☐

1.18 Which one of the countries is known as the land of monsoon?
(a) Russia ☐ (b) India ☐ (c) Australia ☐

1.19 The earth receives major energy from:
(a) Sun ☐ (b) Moon ☐ (c) Mars ☐

1.20 Which one of the following does not contain biomass?
(a) Municipal garbage ☐
(b) Sewage ☐
(c) Metal ☐

2. Answer **only ten** questions (within 2–3 lines): 2 × 10

2.1 What is meant by environmental studies?
2.2 What is succession?
2.3 Define wetland.
2.4 State the function of decomposers in an ecosystem.
2.5 What is biomagnification?
2.6 What is ecosystem?
2.7 Mention the major causes of air pollution in the urban areas.
2.8 What is biogas?
2.9 Define sustainable development.
2.10 State the importance of ozone layer.
2.11 What is Green Bench?
2.12 State the role of bacteria in nitrogen cycle.
2.13 What is meant by non-conventional energy?
2.14 What is exponential growth of population?
2.15 What is fly ash?
2.16 Define carrying capacity.
2.17 What is aquifer?
2.18 Define aerosol.
2.19 What is EIA?
2.20 What is geothermal energy?

3. Answer **any one** question (within 120 words) 6 × 1

3.1 Briefly mention the role of biotic components of an ecosystem.
3.2 State the ecological and economical importance of biodiversity.
3.3 Briefly explain hydrological cycle.
3.4 State how population growth and natural resources are related.
3.5 Explain Gaia hypothesis.
3.6 Give an outline of environmental ethics.

SAMPLE PAPER–III

The figures in the margin indicate full marks

1. Answer **any twelve** questions of the following giving (✓) marks in appropriate box: 2×12

1.1 Seismograph is used for the measurement of
(a) Earthquake ☐ (b) Intensity of light ☐ (c) Humidity ☐

1.2 Jaldapara is famous for the conservation of
(a) Tiger ☐ (b) Rhinoceros ☐ (c) Elephant ☐

1.3 Earth Day is celebrated on
(a) 5th June ☐ (b) 12th May ☐ (c) 22nd April ☐

1.4 The mean temperature of Earth is approximately
(a) 10°C ☐ (b) 15°C ☐ (c) 20°C ☐

1.5 Cholera is a
(a) Vectorborne disease ☐ (b) Waterborne disease ☐
(c) Airborne disease ☐

1.6 India is a megadiversity country with
(a) 4 Ecological hotspots ☐ (b) 3 Ecological hotspots ☐
(c) 2 Ecological hotspots ☐

1.7 The permissible upper limit of Arsenic in water is
(a) 0.05 mg per litre ☐ (b) 0.005 mg per litre ☐
(c) 0.5 mg per litre ☐

1.8 An ecological thinking in which no particular importance is given to man
(a) Social Ecology ☐ (b) Human Ecology ☐
(c) Deep Ecology ☐

1.9 The novel 'Aronnyok' is written by
(a) Rabindranath Tagore ☐ (b) Bibhutibhusan Bandopadhyay ☐
(c) Jibanananda Das ☐

1.10 The historic Earth Summit held at Rio, Brazil in June 1992 is known as
(a) UNCED ☐ (b) UNIDO ☐ (c) UNESCO ☐

1.11 Ramsar Convention includes one of the following
(a) Rabindra Sarovar ☐ (b) Subhas Sarovar ☐
(c) East Kolkata Wetland ☐

1.12 The lowest layer of atmosphere
(a) Ionosphere ☐ (b) Troposphere ☐
(c) Stratosphere ☐

1.13 Bhopal gas disaster caused by the gas
(a) Carbon monoxide ☐ (b) Chlorofluorocarbon ☐
(c) Methyl isocyanate ☐

1.14 'Arabari Model' in West Bengal is famous for
(a) Joint Forest Management ☐ (b) Rainwater Harvesting ☐
(c) Solar Power Plant ☐

1.15 The global human population is around
(a) 480 crores ☐ (b) 620 crores ☐
(c) 760 crores ☐

1.16 The name of Sundarlal Bahuguna is related to
(a) Chipko Movement ☐ (b) Narmada Bachao ☐
(c) Silent Valley ☐

1.17 The famous 3P formula on environment is
(a) Power-Production-Price ☐ (b) Principle of Population Problem ☐
(c) Population-Poverty-Pollution ☐

1.18 On earth the autotropic components can directly fix
(a) Chemical energy ☐ (b) Mechanical energy ☐
(c) Light energy ☐

1.19 The unit of dose of ionising radiation is
(a) ROM ☐ (b) RAD ☐
(c) CAD ☐

1.20 B.O.D. determines the health of
(a) Soil ☐ (b) Air ☐
(c) Water ☐

2. Answer **any ten** questions (with 2–3 lines): 2 × 10

2.1 State two causes of landslide.
2.2 Name two nitrogen fixing bacteria.
2.3 What is an inverted pyramid?
2.4 What is wasteland?

2.5 What is social forestry?
2.6 What is carrying capacity?
2.7 What is photochemical smog?
2.8 What is Decibel?
2.9 What is ozone hole?
2.10 What is urbanisation?
2.11 What is Chipko movement?
2.12 What is bioremediation?
2.13 What is environmental ethics?
2.14 What is bio-diesel?
2.15 What is El Nino?
2.16 What is plate tectonics?
2.17 What is quarantine?
2.18 What is biodiversity?
2.19 What is a cyclone?
2.20 What is acid rain?

3. Answer **any one** question (within 120 words) 6 × 1

3.1 Briefly state the status of air pollution status of Kolkata city.
3.2 Describe the importance of land use planning.
3.3 Briefly state your view about the riverlinking proposals in Indian context.
3.4 State the reasons of wildlife extinction.
3.5 State the impact of economic development on environment in India.
3.6 Discuss your view about the importance of water conservation in our country.

SAMPLE PAPER–IV

The figures in the margin indicate full marks.

1. Answer **any twelve** questions of the following giving (✓) marks in appropriate box: 2×12

1.1 Rachel Carson is the writer of the following book
(a) Biodiversity ☐ (b) Silent Spring ☐
(c) Silent Valley ☐

1.2 5th June is observed as:
(a) Earth Day ☐ (b) Environment Day ☐
(c) Biodiversity Day ☐

1.3 A chronic disease called 'Silicosis' involves:
(a) Heart ☐ (b) Lung ☐
(c) Liver ☐

1.4 A Bengali novel of forestry and environment is
(a) 'Rupashi Bangla' ☐ (b) 'Sabhyatar Sankat' ☐
(c) 'Aronnyok' ☐

1.5 Pollen can cause
(a) Typhoid ☐ (b) Malaria ☐
(c) Allergy ☐

1.6 Free floating microscopic organisms are known as
(a) Nekton ☐ (b) Plankton ☐
(c) Periphyton ☐

1.7 What happened in India in 1984?
(a) Chipko movement ☐ (b) Bhopal gas disaster ☐
(c) Narmada Bachao Andolan. ☐

1.8 Lithosphere represents the earth's
(a) Water ☐ (b) Land ☐
(c) Life ☐

1.9 Chernobyl nuclear power disaster took place in
(a) 1984 ☐ (b) 1986 ☐
(c) 1988 ☐

1.10 *Plasmodium vivax* can cause
(a) Filaria ☐ (b) Malaria ☐
(c) Dengu ☐

1.11 The name of Rabindranath Tagore is connected with
(a) Joint Forest Management ☐ (b) Vanamahatsava ☐
(c) Tree hugging movement ☐

1.12 The physical space occupied by an organism is known as
(a) Niche ☐ (b) Habitat ☐
(c) Ecotone ☐

1.13 Earthquake is measured by
(a) Electroencephalograph ☐ (b) Seismograph ☐
(c) Barometer ☐

1.14 'Study of habitat' is known as
(a) Ecology ☐ (b) Entomology ☐
(c) Ethology ☐

1.15 Aggregation of organisms of different populations refers to
(a) Density ☐ (b) Diversity ☐
(c) Population ☐

1.16 Following is a primary consumer
(a) Green plant ☐ (b) Deer ☐
(c) Tiger ☐

1.17 Biological control refers to
(a) Control of an organism by another organism ☐
(b) Control of an organism by using pesticide ☐
(c) Control of organism by themselves ☐

1.18 Earth is a
(a) Planet ☐ (b) Galaxy ☐
(c) Star ☐

1.19 'Agenda 21' refers to
(a) 1972 Stockholm Summit ☐ (b) 1992 Rio Summit ☐
(c) 2002 Johannesburg Summit ☐

1.20 Following is not found in natural condition in India
(a) Tiger ☐ (b) Zebra ☐
(c) Rhinoceros ☐

2. Answer **any ten** questions (within 2–3 lines): 2 × 10
2.1 What is a Biome?
2.2 What is Green belt?
2.3 What is Minamata disease?
2.4 What is Sanctuary?
2.5 What is Biofertilizer?
2.6 What is Metropolis?
2.7 What is IUCN?
2.8 What is Biological warfare?
2.9 What is organic farming?
2.10 What is Environmentalism?
2.11 What is endangered species?
2.12 What is Dam?
2.13 What is waste?
2.14 What is afforestation?
2.15 What is soil erosion?
2.16 What is Euro II (Bharat Stage II) in auto emission?
2.17 What is an 'Ecocity'?
2.18 What is E.I.A.?
2.19 Name four common waterbirds.
2.20 Name four common water pollutants.

3. Answer **any one** question (within 120 words) 6 × 1
3.1 Briefly state the environmental consciousness of Indian people.
3.2 Make comments on the implementation of environmental laws in India.
3.3 Briefly discuss the different components of a forest ecosystem.
3.4 What are the direct and indirect uses of Biodiversity.
3.5 Discuss the different alternative sources of energy.
3.6 Briefly state the importance of water conservation in our country.

APPENDICES

APPENDIX–1

THE TIME SCALE OF EVOLUTION

Time	*Event*
6.5–7.00 billion years ago	The Big Bang
4.50 billion years ago	The Earth was born
3.50 billion years ago	The blue-green algae (plant) was formed
3.00 billion years ago	The animal appeared
0.0025–0.0030 billion (25–30 lakh) years ago	The man appeared

APPENDIX–2

THE INTERNATIONAL SI UNITS

The modern textbooks have adopted the System International (SI) Units. However, these have not yet been accepted in toto. In this book, some SI units, universally adopted, have been used while the more familiar units, e.g., Å, atm, degree celsius, etc. are retained.

Physical quantity	*Name of Unit*	*Symbol*
Length	metre	m
Mass	kilogram	kg
Amount of substance	mole	mol
Energy	joule	J
Temperature	kelvin (1°C = K – 273)	K

1 Calorie (thermochemical) = 4.184 J

APPENDIX–3

DRINKING WATER QUALITY STANDARDS

In this Table, the permissible limits for water quality parameters, as laid down by the United States Public Health Standard (USPH) for drinking water and Indian Standards Institution (ISI), are listed for comparison. The ISI values, available for only a few parameters, are found to be much higher than those for USPH for no valid reasons. All the units, except for pH and *E. coli*, are in ppm or mg/ml.

Parameters	*USPH Standard*	*ISI Standard*
pH	6.0–8.5	6.0–9.0
Dissolved Oxygen (D.O.)	4.0–6.0	3.0
Chemical Oxygen Demand (COD)	4.0	—
Total dissolved solids	500	—
Suspended solid	5.0	—
Chloride	250	600
Sulphate	250	1000
Cyanide	0.05	—
Nitrate + Nitrite	< 10	—
Fluoride	1.5	3.0
Phosphate	0.1	—
Ammonia	0.5	—
Total hardness (as $CaCO_3$)	500	—
Calcium	100	—
Magnesium	30	—
Arsenic	0.05	0.2
Cadmium	0.01	—
Chromium (VI)	0.05	0.05
Iron (filtrable)	< 0.3	—
Lead	0.05	0.01
Mercury	0.001	—
Zinc	5.5	—
Phenols	0.001	0.005
E. coli (Coliform cells)/1000 ml (Bacteriological parameter)	100	—

APPENDIX–4

TOXIC CHEMICALS IN THE ENVIRONMENT

There are a number of chemicals in the environment. Some of them are toxic while the rest are non-toxic. The toxic chemicals are discharged by industries into air, water and soil. They enter our biological system through food chain and disturb the metabolic processes, leading in some cases to fatal results.

At present there are four million known chemicals in the world and we are adding to the list, 30,000 new compounds every year. Among these 60,000–70,000 chemicals are commonly used. Apart from their benefit to increasing production, living standards and health, many of them are potentially toxic.

Toxic Chemicals in Air

- Acrylonitrile, Arsenic, Asbestos
- Benzene, Beryllium
- Cadmium, Chromium, Chlorinated solvents, Coke oven emissions, Chlorofluorocarbons
- Ethylene dibromide, Ethylene oxide
- Lead, Mercury
- Ozone
- Polycyclic aromatic hydrocarbons
- Sulphur dioxide

Toxic Chemicals in Water

Trace Elements

- Arsenic
- Beryllium, Boron
- Cadmium, Chromium, Copper
- Fluoride
- Lead
- Manganese, Mercury, Molybdenum
- Selenium
- Zinc
- Pesticides (from agricultural run-off)

APPENDIX – 5

THE HAVES AND HAVE-NOTS (Developed and Developing Countries)

80 : 20 Ratio 20 per cent of the world population (developed countries) consume 80 per cent of the world's resources (natural), generate 75 per cent of total solid wastes and are responsible for 70 per cent of global environmental damage.

USA, with about half of India's population and double of India's area, discharges 5000 million tonnes of carbon dioxide every year–five times more than India.

Al Gore, ex-Vice President of USA–UNEP magazine vol. 6, No. 2, 1994–The North (developed countries) accounts for overconsumption of resources while the south (developing countries), for population explosion and poverty.

APPENDIX – 6

ONE EARTH, ONE FAMILY (Global Village)

The UNEP (United Nations Environment Programme) theme for 1994 was *One Earth, One Family*:

- The Earth is not merely a "hotel"–it is our "home".
- it is a place, where we should not use the resources and then move to another spot for further exploration.
- it is a place that we should cherish and need to protect.
- it is not merely a "resource" or "retreat", which we use, enjoy and then leave.
- it is our **Mother.**

APPENDIX – 7

INDIA'S RANK IN THE WORLD

Parameter	*Rank*
Population	2 (next to China)
Area	7
Value added in agriculture	5
Electricity generation	8
Value added in manufacturing	14
Gross Domestic Product	15
Export of goods and services	30
Human Development Index	128

APPENDIX–8

ENVIRONMENTAL SNAPSHOTS

(Our Planet, UNEP Magazine, Vols. 6–7, 1994–95, Nairobi, Kenya)

The Time Scale of Evolution

Time	*Event*
13.76 billion years ago	The Big Bang
4.50 billion years	The earth was born
3.50 billion years ago	The blue-green algae (plant) was formed
3.00 billion years ago	The animal appeared
0.0025–0.0030 billion years ago	The man appeared

The Nuclear Power Plant in the Sky

The sun is the nuclear power plant in the sky. It operates on a 10-billion years supply of hydrogen fuel, producing a steady flow of heat and light and sustains all life on earth. The floods of energy pouring into space from the sun originate in a thermonuclear furnace at its core. Under incredibly intense temperature, pressure and density the fusion reaction: ${}_1^2\text{H} + {}_1^2\text{H} \rightarrow {}_2^4\text{He} + 17\text{MeY}$ occurs with release of tremendous energy in the form of γ-rays and neutrinos. The sun's surface temperature is about 5700 K and core temperature, 15 million K.

This power plant works for the benefit of the entire solar system including the earth.

The Nuclear Tests by Humans

Country	*No. of tests*	*No. of weapons (approx.)*	
USA	1075	(10,000)	
Russia	750	(10,000)	
France	250	(450)	These tests are meant
China	75	(400)	for defence/destruction
U.K.	75	(300)	
India	6	—	
Pakistan	6	—	

The Haves and Have-nots (Developed & Developing Countries)

80:20 Ratio: Twenty per cent of world population (developing countries) consume eighty per cent of world's natural resources, generate seventy-five per cent of all solid wastes and are responsible for seventy per cent of global environmental damage.

USA with about half of Indian population discharges 5000 million tonnes of CO_2 into the atmosphere, while India, 1000 million tonnes of CO_2 every year.

(AI Gore, Vice-President of USA—UNEP magazine, Vol. 6, No. 2, 1994)

The North accounts for overconsumption while the South, for population explosion and poverty.

Environmental Impact Assessment (EIA)

$$I = PAT$$

where I = Impact, P = Population size, A = Affluence (per capita consumption) and T = damage done by technologies chosen to supply each unit of consumption.

The average rich nation citizen used 7.4 kilowatts (kW) of energy in 1990—a continuous flow of energy, equivalent to that powering 74–100 watt light bulbs. By contrast, the average citizen of a poor nation used only 1 kW in 1990. The total environmental impact, as measured by energy use, by the rich nations (1.2 billion people), was 8.9 trillion watts (TW) (7.4 kW × 1.2 billion) while that by the poor nations was 4.1 TW (4.1 billion people).

(Paul Ehrlich—UNEP magazine, Vol. 6, No. 3, 12–13 (1994)).

Penalty for Smoking

In 1970, US Congress passed a law banning cigarette advertisements on radio and TV.

In 1997, tobaccoo giants e.g., Laggard and Brooke were compelled by attorney-generals of 40 US states to pay a fine of $390 billion as compensation over 25 years to meet health-care costs for smoking-related diseases.

The European Community (EC) states clamped a hefty tax 70 per cent on tobacco products. While smoking has abruptly declined in the West, it has registered four-fold increase in India since 1951–52 in consumption of tobacco through cigarettes—21 million kg in 1951–52 to 80 million kg in 1994–95. The government of India has taken the policy of promoting tobacco cultivation. Indian cigarette (1 mg nicotine, 21 mg tar besides HCN, phenol and benzopyrene) is more toxic than an American or European cigarette.

Smoking and Tobacco-related Diseases

According to WHO, tobacco kills one person every 10 seconds and is the singie-largest cause of diseases. By 2020, WHO estimates that tobacco alone will kill 1:5 million Indians. India already accounts for one-third of three million tobacco deaths in the world every year.

Tobacco smoke contains more than 40 chemicals including carcinogens which can, cause cancer, health diseases, respiratory illness, etc. Passive smokers (inhaling tobacco smoke) expose themselves to lung cancer, heart diseases, premature ageing while women smokers are vulnerable to abortion, premature delivery and intrauterine growth retardation.

Smoking costs India a million lives every year and Rs. 14,000 crore in tobacco-related diseases. 31st May every year is observed as Anti-tobacco Day all over the world.

The Endangered Nations

Several hundred beautiful small islands in the Atlantic, Caribbean, Mediterranean, Pacific, Indian Ocean, and South China Sea are facing extinction by rising sea level due to global warming and hence classified as the endangered nations. These islands are located within 2–3 metres above sea level. They are the most vulnerable but least responsible for their risks (global warming).

Thirty-six small island states formed a group and launched "Alliance of Small Island States (AOSIS)" in 1990 at the Second World Climate Conference in Geneva. These include Singapore, Cook Islands, Cyprus, Malta, Maldives, Mauritius, Jamaica, etc.

One Earth, One Family (The Global Village)

For 1994, the UNEP theme was: "One Earth", One Family, The Earth is not merely a *hotel*—it is our *home.*

It is a place, where we should not use its resources and then move to another spot for further exploitation

— It is a place that we should cherish and need to protect.

It is not merely a *resource* or *retreat* (which we use, enjoy and then escape).

— It is our *"Mother".*

The Peace Dividend

About 75 % of the global arms trade ($1 trillion) is directed towards Third World clients and 86 % of these arms are supplied by the five permanent members of the UN Security Council—hardly the role envisaged for the guardians of international security.

Twenty-four developing countries spend over 5 % of their GNP on defence while twelve spend more than 10 % (per capita, p.a. India $9.0, Pakistan $27.0).

If 25 % of the world's defence budget ($1 trillion) could be set aside for 10 years, then some of the key environmental issues facing us today could be solved within this period (Issues—Population, Health, Water & Sanitation, Education, Global Warming, Ozone Hole, etc.).

ENVIRONMENTAL CALENDER

World Wetland Day	Feb. 2
World Forest Day	March 21
World Day for Water	March 22
World Meteorological Day	March 23
Earth Day	April 22
International Biodiversity Day	May 22
Anti-tobacco Day	May 31
World Environment Day	June 5
World Ocean Day	June 8
World Population Day	July 11
Ozone Week	Sept. 16–23
World Car-free Day	Sept. 22
Green Consumer Day	Sept. 28
World farm Animal's Day	Oct. 2
World Habitat Day	Oct. 3
World Animal Welfare Day	Oct. 4
Wildlife Week	Oct. 1–7
World Conservation Day	Oct. 24
International Day for Natural Disaster Reduction	Oct. 13
International Day for Biological Diversity	Dec. 29

❍❍❍

GLOSSARY

Age Structure: Percentage of men and women in the young, adult and old age group in the population.

Acid Rain: Rain water containing mixtures of acids (nitric, hydrochloric and sulphuric acid) from polluted air is known as acid rain. It damages lakes, forests and marble sculptures.

Air Pollution: Various gaseous pollutants from natural and man-made source enter the atmosphere and have adverse effects on the normal properties of air. This leads to air pollution, which is harmful for man and environment.

Air Pollutants: Gases and particles which cause air pollution are called air pollutants.

Air Quality: Each pollutant in air has a limiting concentration, which if exceeded, causes public health hazards.

Aquifer: A highly permeable layer of sediment or rock containing water.

Anthropogenic: Caused by man or man-made things.

Bioaccumulation: Accumulation of non-biodegradable substances in the body (e.g., lead, mercury, DDT, etc.) through food chain.

Biodiversity: Genetic variety among individual species and between species of plants, micro-organism and animals. There are about 10 million species of plants, micro-organisms and animal on earth. These living natural resources are essential for the welfare of mankind.

Biogeochemical Cycles: Circulation of nutrients in cycles among air, soil, water and micro-organisms.

B.O.D.: Biological oxygen demand—amount of dissolved oxygen required by micro-organisms to breakdown organic matter present in water.

Biomass: Organic matter produced by living organisms.

Biome: A broad region-based ecosystem with distinct climate, soil, flora and fauna.

Biosphere: Composite environment consisting of land, air, water, micro-organism, plants, animals and man. Biosphere and environment have close interaction with each other.

Biosphere Reserve: World heritage sites, identified by ICECN, due to their high biodiversity and unique ecosystem e.g., Silent Valley (Kerala).

Carcinogen: Chemicals promoting cancer e.g., benzo-a-pyrene, arsenic, DDT, etc.

Carrying Capacity: Maximum population size that a given system can support over a given period of time.

Chlorofluorocarbons (CFC): Used as solvent, refrigerant, fire retardant, etc. Responsible for ozone hole and greenhouse effect.

Community: Population of various species living and interacting in a given area.

Compost: A nutrient rich soil produced by decomposition of organic matter under aerobic conditions.

DDT: Pesticide useful in agriculture and eradication of malaria.

Doubling Time: Period during which population doubles itself. It is about 100 years in developed countries and 40 years in India.

Dependency Ratio: Ratio of people 65 + (above 65) and 15 (under 15) to the rest of the population.

Ecology: Study of interactions of living organisms with their biotic and abiotic environments.

Ecosystem: A biological community and its physical environment exchanging matter and energy.

Environment: Something that environs i.e., encircles all our surroundings—the natural world in which we live—the living and non-living objects around us in our day-to-day living.

Environmental Studies: The studies of the quality of environment and all aspects of human environment, their degradation, etc. comprise environmental studies.

Eutrophication: Overnourishment of water bodies due to excessive nitrates and phosphates received through run-off—it is harmful for the waterbodies.

Environmental Impact Assessment (EIA): A systematic analysis of the effects of a major development project on environment.

Extinction: Loss of species on earth as natural or man-made process.

Fauna: All the animal species present in a given region.

Flora: All plant species present in a given region.

Food Chain: A feeding series in an ecosystem e.g. plants → animals → man.

Fossil Fuels: Fuels produced by fossilization of plants/ animals such as petroleum, coal, natural gas.

Gasohol: Mixture of gasoline and alcohol is known as gasohol, used as a fuel in Brazil for running cars and buses.

GNP (Gross National Product): An index of a country's economic status based on consumerism i.e., commodities purchased per year, consumer durables and financial status of consumer.

Greenhouse Effect (Global Warming): Rise in temperature of the earth's surface due to increase in the levels of greenhouse gases viz., carbon dioxide, methane etc. The latter trap heat from the earth's surface and returns it thereby raising the earth's surface temperature. This phenomenon is similar to trapping of heat in glass-covered green house (used for growth of vegetables and flowers during winter) and hence called greenhouse effect.

Greenhouse Gases: Gases such as carbon dioxide, methane, water vapour etc. which absorb earth's infra-red radiation, return it to the earth's surface thereby raising the temperature (global warming).

Groundwater: Water held in aquifers 50% below the earth's surface. This is the major source (0.66%) for freshwater.

Heterotroph: Organism that can't synthesize its own food and derives its nourishment by feeding on other.

Humus: A dark amorphous substance that is partially degraded and. serves as a major source of nutrients to plants.

Hurricanes: Cyclonic storms with heavy rains and wind at speed exceeding 120 km/ per hour.

Infant Mortality: Number of infants born per 1000 who die before their first birthday.

Life Expectancy: Average number of years a new-born baby is expected to survive.

Lithosphere: Outer shell of the earth's crust, made of the mantle of rocks. It includes the soil which covers the rock's crust in many places.

Magma: Molten rock below the earth's surface.

Monoculture: Cultivation of a single crop of tree.

Natural Gas: Underground deposits of gases containing mainly methane and small amounts of propane and butane. It is a cleaner fuel than fossil fuel as it produces less carbon dioxide on burning.

Natural Hazards: Hazards from natural sources (earthquake, volcanic eruption etc.) which destroy or damage human lives, houses etc.

Neutron: Nuclear particle with zero charge and mass, (relative to it).

Nitrogen Cycle: Continuous exchange of nitrogen within the ecosystem: air-soil-water.

Nitrogen Fixation: Conversion of atmospheric nitrogen gas into ammonia by nitrogen-fixing bacteria/cyanobacteria.

Nuclear Fission: Splitting of nucleus (uranium/plutonium) into two or more fragments with release of tremendous amount of energy. This is the source of nuclear power (electricity generation).

PAN (Peroxyacetyl Nitrate): A toxic product of photo-chemical smog reactions.

Particulate Matter: Solid particles or ligand droplets suspended in air. Examples are smog particles from combustion of fossil fuels.

Photosynthesis: Synthesis of food (carbohydrates) by green plants in the presence of sunlight using carbon dioxide from air and water from soil.

Photovoltaic Cell: Solar cell that converts, solar energy into electricity.

Population Explosion: Excessive growth of population as in developing countries, to a size that exceeds the carrying capacity.

Primary Pollutants: Gaseous and particulate pollutants discharged directly into the atmosphere by automobile exhaust emissions. The gases thus discharged are carbon monoxide, nitrogen oxides, sulphur dioxide, and hydrocarbons.

Residence Time: Length of time for which a chemical or molecule stays in the environment. For example, the residence time of CFC (chlorofluorocarbon) molecule in the stratosphere is 100 years.

Sanitary Landfill: Waste disposal site on land in which solid waste is spread with fresh layer of clay.

Secondary Pollutant: Harmful pollutants generated from primary pollutants air, for example, sulphonic acid is a secondary pollutant, produced by the primary pollutant sulphur dioxide by reaction with water vapour in air.

Sustainable Development: Improvement in quality of life over a long-term without degrading the environment or compromising the needs of future generations.

Tectonic Plates: Huge blocks of earth's crust which slide along slowly.

Temperature Inversion: The region of increasing temperature with altitude is called temperature inversion. It exists at an altitude 400–600m. below layer of warm air.

Transpiration: Loss of water from plant surfaces.

Troposphere: The bottom region of the atmosphere at an altitude of 0–11 km. It contains 70 per cent of air masses which are always in motion. Here temperature decreases with increasing altitude.

Stratosphere: This region is above the troposphere at an altitude of 11–50 km above the earth's surface. Here ozone acts as a protective shield against ultraviolet radiation from space and thus protects life on earth.

Urbanization: Increasing population in cities by migration from villages and other states. India has a huge urban population (about 300 million).

Waterlogging: Saturation of soil with irrigation water or excessive rain whereby water table rises close to surface.

Watershed: Land area from which water drains under gravity to a common drainage channel.

Wetlands: Ecosystems with stagnant water pool and having rooted vegetation.

❍ ❍ ❍

References

1. A.K. De—*Environmental Chemistry*, 6th edition, New Age International (P) Ltd., New Delhi (2006).
2. A.K. De—*Environmental Studies*, Rural Technology Centre, University of Burdwan (1999), Chapters 1–4.
3. A.K. Agarwal (Editor)—*The State of India's Environment*, (1984–85), Centre for Science and Environment, New Delhi (1985).
4. A.K. Mitra—*Urban Environment in Crisis*, New Age International (P) Ltd., New Delhi (2000).
5. A.K. De—*Environmental Studies*, 2nd edition, New Age International (P) Ltd., New Delhi (2005).
6. A.K. De—*Environmental Engineering*, New Age International (P) Ltd., New Delhi (2009).
7. D.D. Monte—*Temples or Tombs*? Centre for Science and Environment, New Delhi (1985).
8. G.D. Friedman—*Primer of Epidemiology*, 3rd ed., McGraw-Hill, New York (1988).
9. T.M. Das—*Man, an Endangered Species*, Shaibya Prakasan, Kolkata (1995).
10. T.M. Das—*Is the Earth for Man Only*? Jonaki Publishers, Kolkata, 4th edition, (1995).
11. T.M. Khosoo—*Environmental Priorities in India and Sustainable Development*, Indian Science Congress Association, New Delhi (1986).

INDEX